Public Understanding of Science

Between the French Revolution in 1789 and the 'chemists' war' (1914–18), science became culturally and economically vital; seemingly pervasive but also difficult. This book explores how science was disseminated during this period, moving from a time in the late eighteenth century when science was not widely regarded as a necessary tool for investigating the world to the start of the twentieth century, when it was crucial.

In the era of political and industrial revolution, preachers, poets, artists, writers and lecturer–performers attracted large publics ready to be convinced of intellectual and social progress made visible through science. Did scientists (a nineteenth-century word) have an easily learned and more widely applicable method? Who was best at communicating it: scientists, popularisers or critics? David Knight's fascinating history reveals how the successes and failures of our ancestors help us understand the position science comes to occupy now.

Ruefully, scientists see that 'chemical' is a dirty word, and technology is criticised, whilst 'alternative' therapies flourish. We now live in a culture of suspicion, where experts are distrusted. Seeing how enthusiasm for science was kindled in the nineteenth-century Age of Science casts light on our current situation. This highly engaging, readable book will be of great interest to scientists and professionals alike, as well as to literary critics and historians; it will be equally accessible to the general reader.

David Knight is Emeritus Professor of History and Philosophy of Science at the University of Durham. His distinguished career has led to numerous awards including a Templeton Foundation award for teaching 'Science & Religion in the 19th century', and the American Chemical Society's Edelstein Award for History of Chemistry. His most recent publications include *Science and Beliefs: From Natural Philosophy to Natural Science, 1700–1900* (co-editor), and *Science and Spirituality: the Volatile Connection* (Routledge, 2004).

Routledge studies in the history of science, technology and medicine
Edited by John Krige
Georgia Institute of Technology, Atlanta, USA

Routledge Studies in the History of Science, Technology and Medicine aims to stimulate research in the field, concentrating on the twentieth century. It seeks to contribute to our understanding of science, technology and medicine as they are embedded in society, exploring the links between the subjects on the one hand and the cultural, economic, political and institutional contexts of their genesis and development on the other. Within this framework, and while not favouring any particular methodological approach, the series welcomes studies which examine relations between science, technology, medicine and society in new ways, e.g. the social construction of technologies, large technical systems, etc.

1. **Technological Change**
   Methods and themes in the history of technology
   Edited by Robert Fox

2. **Technology Transfer Out of Germany after 1945**
   Edited by Matthias Judt and Burghard Ciesla

3. **Entomology, Ecology and Agriculture**
   The making of scientific careers in North America, 1885–1985
   Paolo Palladino

4. **The Historiography of Contemporary Science and Technology**
   Edited by Thomas Söderquist

5. **Science and Spectacle**
   The work of Jodrell Bank in post-war British culture
   Jon Agar

6. **Molecularizing Biology and Medicine**
   New practices and alliances, 1910s–1970s
   Edited by Soraya de Chadarevian and Harmke Kamminga

7 **Cold War, Hot Science**
Applied research in Britain's defence laboratories 1945–1990
Edited by Robert Bud and Philip Gammett

8 **Planning Armageddon**
Britain, the United States and the command of Western Nuclear Forces 1945–1964
Stephen Twigge and Len Scott

9 **Cultures of Control**
Edited by Miriam R. Levin

10 **Science, Cold War and the American State**
Lloyd V. Berkner and the balance of professional ideals
Alan A. Needell

11 **Reconsidering Sputnik**
Forty years since the Soviet satellite
Edited by Roger D. Launius

12 **Crossing Boundaries, Building Bridges**
Comparing the history of women engineers, 1870s–1990s
Edited by Annie Canel, Ruth Oldenziel and Karin Zachmann

13 **Changing Images in Mathematics**
From the French Revolution to the new millennium
Edited by Umberto Bottazzini and Amy Dahan Dalmedico

14 **Heredity and Infection**
The history of disease transmission
Edited by Jean-Paul Gaudilliere and Llana Löwy

15 **The Analogue Alternative**
The electric analogue computer in Britain and the USA, 1930–1975
James S. Small

16 **Instruments, Travel and Science**
Itineraries of precision from the seventeenth to the twentieth century
Edited by Marie-Noëlle Bourguet, Christian Licoppe and H. Otto Sibum

17 **The Fight Against Cancer**
France, 1890–1940
Patrice Pinell

18 **Collaboration in the Pharmaceutical Industry**
Changing relationships in Britain and France, 1935–1965
Viviane Quirke

19 **Classical Genetic Research and its Legacy**
The mapping cultures of twentieth-century genetics
Edited by Hans-Jörg Rheinberger and Jean-Paul Gaudillière

20 **From Molecular Genetics to Genomics**
The mapping cultures of twentieth-century genetics
Edited by Jean-Paul Gaudillière and Hans-Jörg Rheinberger

21 **Interferon**
The science and selling of a miracle drug
Toine Pieters

22 **Measurement and Statistics in Science and Technology**
1930 to the present
Benoît Godin

23 **The Historiography of Science, Technology and Medicine**
Writing recent science
Edited by Ron Doel and Thomas Söderqvist

24 **International Science Between the World Wars**
The case of genetics
Nikolai Krementsov

25 **The Social Construction of Disease**
From Scrapie to Prion
Kiheung Kim

26 **Public Understanding of Science**
A history of communicating scientific ideas
David Knight

Also published by Routledge in hardback and paperback:

Science and Ideology
A comparative history
Mark Walker

Public Understanding of Science

A history of communicating scientific ideas

David Knight

First published 2006
by Routledge
2 Park Square, Milton Park, Abingdon, Oxon OX14 4RN

Simultaneously published in the USA and Canada
by Routledge
270 Madison Ave, New York, NY 10016

Routledge is an imprint of the Taylor & Francis Group, an informa business

© 2006 David Knight

Typeset in Times by Wearset Ltd, Boldon, Tyne and Wear
Printed and bound in Great Britain by TJI Digital, Padstow, Cornwall

All rights reserved. No part of this book may be reprinted or reproduced or utilised in any form or by any electronic, mechanical, or other means, now known or hereafter invented, including photocopying and recording, or in any information storage or retrieval system, without permission in writing from the publishers.

British Library Cataloguing in Publication Data
A catalogue record for this book is available from the British Library

Library of Congress Cataloging in Publication Data
A catalog record for this book has been requested

ISBN10: 0-415-20638-3 (hbk)
ISBN10: 0-203-96642-2 (ebk)

ISBN13: 978-0-415-20638-9 (hbk)
ISBN13: 978-0-203-96642-6 (ebk)

Contents

	Acknowledgements	viii
1	Understanding	1
2	God's clockworld	13
3	Holding forth	29
4	Poetry, metaphor and algebra	44
5	Picturing science	62
6	Ballyhoo	76
7	Display	91
8	Travel	106
9	Imagining	119
10	Science gossip	135
11	Suspending judgement	153
12	Classical physics	167
13	Promoters and popularisers	182
	Notes	197
	Index	228

Acknowledgements

This book could not have been written without help from librarians, especially in Durham University and at the Royal Institution in London. I am grateful to my colleagues in Durham, for support during a term of research leave during which the book was begun, and then subsequently during my process of retirement when it was completed. I am also grateful to friends and colleagues who have listened to papers at conferences and seminars on history of science, from which chapters here have evolved; to Durham University for appointing me to chair its ethics committee for several years; and to the Royal Society of Chemistry, which has elected me to its Committee for the Promotion of Chemistry to the Public – both of these being most instructive experiences. Students in Durham met much of this book as a course on the emergence of two cultures, and I am grateful for their reactions in tutorials.

The author and publishers would like to thank the following for granting permission to reproduce material in this work:

Interdisciplinary Science Reviews for permission to reprint a version of Chapter 4, 'Poetry, metaphor and algebra', previously published as 'Humphry Davy the Poet', by David Knight in *Interdisciplinary Science Reviews*, December 2005 (Vol. 30, number 4, pp. 356–72).

Every effort has been made to contact copyright holders for their permission to reprint material in this book. The publishers would be grateful to hear from any copyright holder who is not here acknowledged and will undertake to rectify any errors or omissions in future editions of this book.

1 Understanding

Science is not easy to understand. And sometimes it is like an ice-dance show: at first you wonder how it's done, and then you wonder why. It is not obvious why people devote themselves to this activity, or why others should be interested and should directly or indirectly pay the bills. This book is about how people have tried to get natural science (physics, chemistry, biology) loved – or sometimes hated, for just as we all find out about religion not only from bishops, rabbis or ayatollahs, so with science the heretics and opponents can teach us as much as the academicians and professors. Mainstream 'popular science' often implies a reductive scientism,[1] the notion that real explanations of anything must always be scientific in the way physics is, and a certain condescension to meaner intellects. It is also full of 'breakthroughs', good for promoting funding and careers, but often of little lasting significance: promises of a world where science and technology will abolish poverty and pain were new and exciting when Humphry Davy made them in his inaugural lecture in London in 1802,[2] but have been often broken since. Just as churches have not always practised the love they preach, so peace and plenty have not always resulted from the open-minded search for truth prominent in scientists' sermons. Mavericks too, promising what contemporaries said was impossible, have always been around to damn the complacent scientific establishment.[3]

Science wars are thus not new. There have always been critics, for some of whom scientists have been comically absent-minded intellectuals, while for others they have been sinister Dr Strangeloves: and polemic against science is as instructive as propaganda for it, and equally enthusiastic. Critics may have some vision of a pre-scientific world we have lost, an Eden, Merrie England or Jeffersonian rural America; or want to leave room for religious or other beliefs, and for humanities, in what seems a cold inanimate world; or may have in mind a scientific world-view, or 'paradigm', different from the current one. Thus two rather dissimilar people, both of them in their different ways enthusiasts for science, rejected mechanical, clockwork analogies widely held around 1800 in favour of a world of forces, with dynamic rather than inert matter: Michael Faraday proposed field theory[4] in place of atoms and action at a distance; and the poet Samuel Taylor Coleridge, while excited by the chemistry of his friend Davy,[5] rejected its claims to account for the processes of life.[6]

We should remember also that much science is rather dull, as those who have studied it formally will know. Doing experiments like putting pennies into concentrated nitric acid and watching the brown fumes, or making gas jars full of hydrogen go pop with a lighted splint, is fun; and so is squirting other people with wash-bottles. But making accurate measurements, weighing things and working steadily through analytical procedures is unexciting; and learning much chemistry is painful, with hard names, complex formulae, and equations difficult to balance. Just so, thinking up hypotheses is fun, but processes of confirmation or refutation are slow, laborious and often involve statistics. That is hard to popularise. Proof is a burden. The science that gets picked up is the glamorous, the sublime perhaps or the manifestly practical; or it is the controversial.

Science has long been associated with great rows.[7] Isaac Newton quarrelled with some of his notable contemporaries, Robert Hooke, John Flamsteed and Gottfried Wilhelm Leibniz. In science, priority really matters: it is a race in which there are no silver or bronze medals. In the nineteenth century, the two leading men in any particular field were very frequently at daggers drawn, with their disciples or scientific children being dragged in like young Capulets or Montagues. Thus in geology, Adam Sedgwick and Roderick Murchison fell out over whether Sedgwick's Cambrian rocks were really just the lower part of Murchison's Silurian; in zoology, Richard Owen and Thomas Henry Huxley quarrelled publicly over topics like the anatomy of the gorilla's brain (and its similarity to ours); and in chemistry, James Dewar and William Ramsay were not on speaking terms, and so Ramsay had to reinvent apparatus suitable for liquefying gases when he isolated argon, neon, krypton and the other 'noble gases' from the atmosphere.[8] It was no better in France, Germany or the USA; and not entirely different in today's bigger scientific world. Public disputes make science a good spectator-sport. There has thus often been an entry into current science through personalities and issues, much more exciting than colder and more formal routes; and there are always scientists who also enjoy playing to a gallery, though others tut-tut at dirty washing being done in public, rather than questions being resolved among experts.

Popular science at any time is therefore for many reasons different from the established kind; and, to the frustration of scientists, public understanding (just like public interest) is very different from theirs. There is no one 'public' after all. Thus there are specialists in different disciplines who want to keep up across the board. They may be supportive of colleagues, or may feel that some sciences are grossly and unfairly over-funded compared to theirs. Then there are other highly-educated people, in humanities, languages, law or social science, 'erudite non-specialists' they have been called, who are again thrilled, intrigued or horrified at what they perceive going on in science. These may be policy-makers, journalists, legislators and others whose opinions are directly important to scientists, and affect their lives. Distinct from these are ordinary people, busy, more or less curious about new ideas or enthusiastic about technical developments, probably suspicious about what supposed experts tell them, and wary of change. There are consumers to be stimulated by scientific-looking advertisements to buy beauty products or pep pills – or to avoid 'chemicals' in the name of nature and the organic. Finally, there are children, the rising generation, whose inquisitive enthusiasm must be maintained if science is to go on. Museums used to be aimed at the more earnest of such publics: nowadays having a good time is more crucial, and 'things', historic exhibits that can give a wonderful feel for the development of the sciences and engineering, may be put in storage so that visitors can play interactive computer games. The balance is not obvious; and to appeal to several publics at once is and was problematic. It matters, because of the place that science holds in our culture and our economies.

The problems became apparent in the long nineteenth century, the 'Age of Science' (when it came to maturity[9]) from the French Revolution of 1789, spurred on supposedly by the intellectuals of the Enlightenment, to the outbreak of the First World War ('the chemists' war') in 1914. The classic work on this period, published a century ago, was John Theodore Merz's *History of European Thought in the Nineteenth Century*, of which the first two volumes were devoted to science.[10] We are all in debt to this German-born and trained electrical engineer who made his fortune in Newcastle. His book was thematic, getting inside the minds of scientists in various traditions: so is mine, but concerned with the outside – looking at various ways in which science was made available to various publics, and sometimes to everyone: in sermons, lectures, verses, pictures, ballyhoo, displays, travellers' tales, journalism at various levels, and then later in the century from newly professional biologists and physicists – and professional popularisers. Some of these attempts to improve understanding were solemn, but in many the aim was to make it fun; and this book will fail if it is not 'entertaining knowledge' also. It is serious too: what was perceived or even generally supposed is as interesting and important, after all, as what happened among insiders.

The nineteenth century saw, in physiology, chemistry, geology and thermodynamics, as well as in technology, the delayed triumphs of the scientific revolution of Galileo and Newton; but also the emergence of experts who no longer shared one common culture – professionally trained scientists, engineers, doctors, nurses, architects, accountants, lawyers, clergy, and even writers and journalists. Renaissance science had been introduced into Europe through contacts with Islam and China: modern science, which became essentially an activity for comfortably off European males, was then opened up during the nineteenth century. It became a route to social mobility. Women's activity, and that of assistants at home, and in the field or the outback, became less covert. By 1900 skilled practitioners were also emerging in India and Japan, and the USA was on the way to becoming a scientific superpower.

In late-eighteenth-century France it was possible to contemplate a career in science, culminating in salaried membership of the Academy of Sciences,[11] though only very few could hope to achieve it. France, meaning Paris, remained the world's centre of excellence in science right through the Revolution, the Terror, the Directory, the Consulate, Napoleon's Empire, and the subsequent Bourbon Restoration after the Battle of Waterloo in 1815 and the Congress of

Vienna that followed it. The Academy of Sciences was briefly closed down as elitist but soon restored, mildly purged, because it was useful – advising on projects such as the melting of church bells into guns. The eminent chemist Antoine Francois Fourcroy made a great name for himself as a public lecturer,[12] attracting enormous audiences at a time when it was politically necessary for science to be made accessible; while at the Jardin des Plantes and its associated museum and zoo, Georges Cuvier and others also gave public lectures on natural history.[13] Science in the capital was effectively popularised by academicians, experts who certainly did not at that time lose kudos by undertaking such tasks. It was the duty of the natural philosopher to communicate his knowledge and world-view as widely as possible; certainly beyond the narrow confines of the scientific community.

In Britain, undergoing the Industrial Revolution which made possible the victory over France, science was more like a hobby than a career. Water frames, spinning jennies and even steam engines did not draw much upon recent or recondite science. Indeed, technological achievements gave rise to scientific problems. The steam engine did more for science than science did for the steam engine: thermodynamics, the study of heat and work, was born from scientific and mathematical analyses of engines, ideal and actual; and the engineers like Thomas Newcomen, James Watt, Richard Trevithick and George Stephenson who built stationary and then locomotive engines had little formal modern science to guide them. Men of science were often doctors like Thomas Garnett and Thomas Young, clergymen like Gilbert White and Joseph Priestley, lawyers like William Grove, or leisured gentry like Henry Cavendish.

Davy, who abandoned a medical training for chemical research and lecturing, was one of the very few who could make a living from science; and even he, like Dick Whittington, completed his social mobility by marrying money. His attention was called to coal mining by a disastrous explosion near Sunderland. Davy's safety lamp for coal miners, resulting in 1815 from a rapid series of experiments done on the explosive 'fire damp' (methane, our CH_4) in the laboratory of the Royal Institution, was one of the very first examples of 'applied science'. The device invented by the genius in the metropolis worked down the pit, saving lives and making possible the expansion of the coal industry that fuelled the British economy right through the Victorian period and beyond.[14] Davy became Sir Humphry, but his rich marriage was childless – there was no son to inherit his title. His lectures, a sensational success in Regency London, and then his practical discovery, were important in getting science across: useful, entertaining, exciting and now also British.

A lamp not unlike Davy's had been made by classic trial and error by George Stephenson, but the publicity machine of the Royal Society, the Royal Institution and the metropolis generally, was very effective. Men of science in Davy's time were called 'natural philosophers', or 'philosophers' for short: their way was what Davy called 'refined common sense' (and what Huxley was to call 'trained and organised common sense') in approaching problems systematically and from first principles.[15] Thus Davy had identified the fire-damp chemically, and explored its properties: this was contrasted to the unenlightened commonsensical approach of practical men involved in the Industrial Revolution. Davy, it was claimed, had come to his lamp 'philosophically'. Such triumphs of applied science were what he had promised in his inaugural lecture, seeing the dawn of a bright day of high technology – in accordance with Francis Bacon's dictum, 'knowledge is power'. By 1815 laboratory science was becoming ever more recondite. If it was beginning to deliver benefits, populisers could emphasise that aspect as they sought public interest and acclaim, and funds to make careers. Scientists duly achieved honour and respect; though some could still be viewed as absent-minded and dotty professors, and others as threatening.

Davy, Faraday (his assistant and great discovery), Fourcroy and Cuvier were great men getting across their own work and that of their peers. As natural philosophers, that was the right thing to do; and the snobbery about showmanship that downgraded the writing of textbooks or popularisations, and the giving of public lectures, came only at the end of the nineteenth century, and in the twentieth.[16] The line between experts and popularisers was fuzzier than it later became: the scientific community was very small, and papers published by the Royal Society in its *Philosophical Transactions* about 1800 were discursive and accessible compared to those appearing a century later. The Fellows of the Society, whose subscriptions kept it afloat, were chiefly landed or professional men (no women until after the Second World War) interested in science, but not active in research or teaching. A minority even of the governing Council had ever published a scientific paper until after Davy became President in 1820; and it was another generation before the Society began to look more like an Academy, composed exclusively of distinguished discoverers. Addressing such a body before about 1850 was not so very different from writing or lecturing for the general well-educated public.

Just as literature had its Grub Street hacks, lice on the locks of literature, so there were some who made their living out of getting science across. Thus Jeremiah Joyce,[17] an ardent radical arrested and charged with high treason in the jittery year 1794, was a Unitarian minister among whose many publications was *Scientific Dialogues*, 1807: a favourite of the young John Stuart Mill,[18] who never remembered 'being so wrapped up in any book', until his formidable father found out, and drew attention to its errors. Reginald Heber, Bishop of Calcutta in the 1820s, found it in use in a regimental school for English and Indian boys in Cawnpoor, but 'the native boys ... had [it] for their single class-book, which they stammered over by rote, but could none of them construe into Hindostanee'. It was odd that a book written to make science intelligible should have been thus incomprehensible. Joyce also published his lectures on science, and many other works, including an updated edition of William Paley's celebrated *Natural Theology*; some of his writings were pseudonymous.

Mill was recommended instead to the writings of his father's early friend and schoolfellow, Thomas Thomson, whose position in the world of science, and target public, were rather different. In his standard textbook of chemistry (which received the accolade of translation into French, the language of Lavoisier), he first made known the atomic theory of John Dalton. He did experiments confirming it, and he ended his career as Professor of Chemistry at Glasgow University, where he made his students do practical work in the laboratory.[19] But there were other popular writers of higher repute than Joyce: Samuel Parkes, an industrial chemist in East London, wrote a very popular *Chemical Catechism*; and the *Conversations on Chemistry* of Jane Marcet, wife of a prominent doctor, also first published in 1807, was specifically aimed at girls, though young Faraday became its most eminent reader.[20]

William Nicholson was more like Thomson in reputation, author of both a dictionary and a textbook of chemistry, and also a translator of scientific works from the French, notably Fourcroy. When Alessandro Volta's paper, in French, announcing the invention of his 'pile' (the ancestor of our electric batteries), was sent to the Royal Society in 1799, Nicholson and the surgeon Anthony Carlisle read it before publication and repeated the experiment. They then extended it, dipping the wires from the terminals into water and observing that hydrogen and oxygen came bubbling up. They were thus among the founders of electrochemistry. Meanwhile, Nicholson had started a *Journal of Natural Philosophy, Chemistry and the Arts* ('arts' meaning 'techniques') in 1797, which ran until 1813, in which year Thomson began his *Annals of Philosophy*, which continued until 1826. Both were swallowed by the *Philosophical Magazine*, edited by Alexander Tilloch, which eventually played an important part in the history of the company, Taylor and Francis, that became its publishers, and which still continues.[21] Tilloch's importance is as an editor and publisher, rather than for any science of his own; and such people have always been of immense importance in the dissemination of science, public knowledge that requires writers, readers, editors and entrepreneurs commissioning and publishing.[22]

These three private journals were aimed at a general readership. Like the Royal Society's more august publication (on excellent paper in quarto size), their crowded octavo volumes covered the whole of science, or most of it, since natural history already had its own vehicles. They included original papers, sometimes important ones, with reprints, news, reviews and correspondence, building up and encouraging a community of readers; and they came out quarterly, without much worry about peer review, thus promising rapid publication to ensure priority.

By 1900, everything was very different. Much more science was known. It had taken Priestley, or later Davy, a few months to pick up enough knowledge to work at the very frontier of knowledge. That was impossible in the days of J.J. Thomson, Marie Curie and Max Planck. Steady development in scientific education, notably from Germany where Justus Liebig at Giessen had invented the research student, working for a PhD, had led to a big and specialised scientific community. The Prussian victories over Austria and then, in 1870, over France had immensely stimulated education in Britain and elsewhere. The perception was that, in 1870, the more educated nation had beaten what had been supposed a much stronger military power. Elementary schooling became compulsory in backward England, and new 'redbrick' universities began to attract students, many of whom took degrees in science or engineering. Universities became centres of scientific research, and the old idea that all students should receive the same basic liberal education (in classics, with maybe mathematics) disappeared. By 1900, there was a network of polytechnics, based upon the German Technisches Hochschulen and ultimately upon the elite Parisian École Polytechnique. Industry, which in Britain especially had been suspicious of book-learning rather than experience, and had employed scientists as consultants when something went wrong, had (in a trend beginning in Germany) become an important employer.[23]

If Charles Snow was right in diagnosing 'two cultures', mutually incomprehensible and perhaps antagonistic, in Britain in the mid-twentieth century (a scientific and a humanistic one), then this situation had been coming about since the later nineteenth century.[24] He found that music was the favourite art of scientists: maybe that was true for the twentieth century, but in the nineteenth it was not obviously the case. We shall be looking at scientific and technical illustration, handsome as well as informative. Faraday admired the visual arts, John Herschel and Henry de la Beche were adepts with a pencil, and the wealthy and cantankerous astronomer Richard Sheepshanks made a wonderful collection of modern paintings, which he bequeathed to the Victoria and Albert Museum. Similarly, Richard Owen, Thomas Huxley and William Clifford loved poetry, and Davy, Herschel and Maxwell wrote it.

We may also doubt Snow's analysis into only two cultures, knowing that chemists and physicists often glowered at each other across a social and intellectual frontier, that historians and literary critics often had little to say to each other, and that social scientists were out on their own. But that only goes to show how fragmented the world of knowledge had become by 1914. Being a 'Renaissance man' was no longer possible. The educated gentleman could not, as Aristotle had hoped, know enough to be able to judge what experts were up to. The brothers Willhelm and Alexander von Humboldt between them knew and contributed to most branches of the knowledge of the early nineteenth century; but by the early twentieth, even such a talented pair could not have done it. Willy-nilly, the world had become specialised: people knew more and more about less and less. Snow was in a line from distinguished Victorians, including William Whewell, the know-all Master of Trinity College, Cambridge, (science his forte, omniscience his foible) who had deplored this trend but could not stop it. Liberal education and common culture were in jeopardy.

The new universities, and even the old ones, admitted women, who could at last come to play a full and public part in science, as they long had in its translation and popularisation: so that the term 'man of science' that had by 1870 generally superseded 'natural philosopher' was in its turn replaced by 'scientist', which Whewell had coined in 1833 by analogy with 'artist', but was slow to gain favour. Spending a life in science, which had been an odd thing to think of in 1800, had become by 1900 a respectable and plausible ambition. Science was, as it had not been in William Blake's day, 'considered to be an inevitable or even necessary way of investigating and understanding the world'.[25] Culture and the economy depended upon it, and ignorance or distaste was beginning to seem shameful.

In 1904 the British Association for the Advancement of Science met in Cambridge, and the President was the Prime Minister, the Conservative Arthur Balfour, a wealthy philosopher and aesthete. He gave an interesting address, calling attention to the great intellectual revolution going on in physics which confirmed his view that science, like everything else, rested upon metaphysical beliefs that were not directly testable. He had been well briefed, but was a man who took a genuine and keen interest in science, and was in the 1920s to become in effect Britain's first Minister for Science, having declined to be nominated as President of the Royal Society. That such a central figure in the establishment should preside at such an occasion was a sign of the full acceptance of the cultural as well as economic importance of science.

Lord Rayleigh, Thomson's predecessor at the Cavendish Laboratory at Cambridge, distinguished for his precision in measurement, had coached Balfour; but most people lacked such distinguished relatives, and were not in Balfour's exalted position. They might hear lectures about science, now illuminated by electricity rather than gas, at the annual meetings of the B.A.A.S., at the Royal Institution in London's West End, or at one of the Literary and Philosophical Societies, Athenaeums, Museums, or Mechanics' Institutes elsewhere. These would be directed at an interested audience, perhaps a well-off and well-educated one, but perhaps composed of working men (a group Huxley particularly liked to address). One aspect of specialisation was that scientists often did not know what their fellows were up to. High-level popularisation, haute vulgarisation, was required to keep physicists aware of developments in biology and so on. Faraday introduced 'Friday Evening Discourses' at the Royal Institution to achieve this end, and those invited to lecture took the task very seriously and saw it as an honour and obligation.

The process had begun, surprisingly enough, with a woman, Mary Somerville,[26] writing books in the 1820s and 1830s which were very well-received by men of science like Whewell and Faraday. Later, William Crookes, the last non-graduate President of the Royal Society (during the First World War) launched a *Quarterly Journal of Science* (in 1864) which he hoped would catch on as the great literary reviews (the *Edinburgh*, the *Quarterly*, the *Westminster*, the *North British* and others) had done. These published essentially essay-reviews, more or less tightly focused upon one or more recent publications; and served to keep nineteenth-century readers up-to-date on a wide range of topics, including some science. Crookes' *Quarterly Journal* was to be distinctive in dealing with the whole gamut of the sciences only. In the event, by 1864 the era of the great quarterlies was coming to an end, and more frequent publication was becoming fashionable: the journal eventually went monthly, but ran only until 1885. Much more successful was Crookes' *Chemical News*, in magazine format and coming out weekly: much more lively than the publications of the Chemical Society of London, informal, often (like *The Lancet* then, in the medical world) critical of the scientific powers-that-be, publishing much speculative and unconfirmed material, and representing a chemical community now containing many 'professionals', living by their science in industry or in government, and remote from the learned world of academic research.

Chemical News was the model for Norman Lockyer's journal *Nature*, published from 1869 by Macmillan, and for many years in the red, supported as a loss-leader. It cast a glow over Macmillan's textbooks and other scientific works, making the publisher a leader in this field, booming with educational expansion. *Nature* was a vehicle for announcements and preliminary papers, often in the form of letters to the editor; and also functioned through reviews to make specialised scientists aware of what was going on elsewhere. It became essential. Like Crookes, Lockyer was an entrepreneur, a self-made man, a civil servant passionate about astronomy and prepared to speculate, about helium, which he identified in the Sun long before it was isolated upon Earth, and about the life history of stars. Both men were good communicators, and made themselves among the best-known scientists of the years around 1900: busy men, they did much of their research by directing assistants. They filled the gap between elite and popular science, and were prepared to delight in argument, public excitement and controversy.

Elsewhere, magazines like *The Cornhill* and *The Nineteenth Century* had brought a liveliness not always visible in the austere pages of the *Reviews*; and *The Nineteenth Century* particularly, whose editor James Knowles was a friend of Huxley and his X-club associates,[27] carried a good deal of science, especially when it could be made controversial: Huxley's exchanges with W.E. Gladstone over evolution and the Bible are an example; and Huxley died in 1895 in the midst of writing an essay engaging with Balfour's view that science, like religion, rested upon belief.[28] These intellectual encounters with two Prime Ministers tells us something about both science and politics at the time. Huxley was a great stylist, whose writings can still be read with pleasure. He had had to learn to write attractively to support himself when, after returning from a survey voyage in Australian waters, he was elected FRS but was unable for some years to find a job. Scientific journalism filled the gap.

Few scientists by 1900 had the common touch of Huxley, Crookes and Lockyer, all in their way plebeians; and Lavoisier's hopes for an austere language of science, where metaphor would be excluded, was to a great extent realised. The passive voice, the long words, and the compressed style required by editors publishing for expert readers made science hard to read. A career in scientific popularisation had opened up, and interpreters of science found their niche. There was also science fiction: *Frankenstein* had been an early example, drawing upon Davy; and then the novels of Jules Verne; and by the end of the century H.G. Wells' *Time Machine*. Wells had briefly been a pupil of Huxley's, and his book explored the themes of evolution and degeneration: like *Frankenstein*'s, his message was not quite what optimistic and progressive boosters of science would have wished.

Similarly, while cheap books and then newspapers had helped popularise mainstream science, they had also brought before the public a series of scientific heretics. 'Scriptural Geologists' interpreted *Genesis* literally, to the horror of liberal Christians and professional geologists; vaccination, welcomed by most as abolishing smallpox and holding out the promise of similar developments with other infections, was violently attacked by some in the nineteenth century, just as the MMR vaccine was in the twentieth; and vivisection, claimed by physiologists to be necessary for medical progress, was furiously denounced, and in the event restricted and controlled by Act of Parliament. Industrial pollution and the adulteration of food aroused similar outcries, and were eventually controlled (creating jobs for analysts). Scientists were also beginning to play the role of sages, or high priests in what Huxley called the Church Scientific, controlling science: churches and nationalised industries seem to be run for the benefit of their officials and employees, and some had the same unworthy suspicion about the scientific community and its establishment. All publicity, we are told in the world of theatre, is good publicity; and maybe in science that will also apply, but scientists like to be loved.

Nevertheless, it was not clear what was and was not science. Phrenologists aroused much attention with lectures, publications and labelled china heads, connecting the shape of skulls with the development of brain, and hence of personality.[29] In 1850 William Gregory, Professor of Chemistry at Edinburgh, who had translated important books by Justus von Liebig, published Karl von Reichenbach's *Researches*, covering magnetism, electricity, heat light, crystals, chemical affinity and the vital force.[30] A chemical engineer by training, Reichenbach had come to perceive magnetic auras around people, which he attributed to a previously unidentified force called odyle, or od for short. Even in the 1920s, this idea, presented as an insight like others in history unappreciated by blinkered experts, was still worth re-publishing by a London publisher.[31] Meanwhile there had been the great excitement caused by the coming of spiritualism, and its seances. In this case, the Society for Psychical Research, including eminent scientists, was founded to investigate the curious phenomena observed in the presence of respectable witnesses.[32] They had been literally in the dark; and those trying to investigate and explain found themselves metaphorically there. But telepathy, disappointing though it was in its failure to be dependable and reproducible, was studied, and the unconscious postulated. Science for the public was not straightforward, and the favourites of some publics were disapproved of by leading scientists – though other unrespectable ideas, like evolution, which had been less well-thought-of than phrenology, came triumphantly through expert hostility to flourish in our time.

The nineteenth century was an age of empires and colonies, the settling of Australasia and the American West, the consolidation of British India and French Algeria, and the scramble for Africa. All this generated and required science.[33] In 1800 much of the globe was still blank. 'Darkest Africa' became proverbial, but the interior really was unknown then to Europeans. James Cook had made it clear that there was no great unknown temperate southern continent, but whether New South Wales and New Holland formed one land mass or were a collection of islands was unclear, and the interior of what Cook's successor, Matthew Flinders, was to call Australia was also unknown. Central Asia was a mystery. South and Central America too were little explored, and while Alexander Mackenzie had reached the Pacific in Canada,[34] nobody had yet crossed the Rocky Mountains in what became the USA; California was still Spanish. When, as a student in 1959, I went to Madagascar, there were still blank spaces on the map, great tracts of forest without air or ground survey; but really by 1900 there were very few areas quite unknown to geographers. The world had been explored. This was partly due to government-sponsored expeditions, 'big science', like Cook's, from Britain, France, the USA and Russia in particular; and partly to much smaller and cheaper journeys by individuals or small groups, like Mackenzie's had been. Sailing ships, with all the skills involved and the problems they presented on lee shores, in calms, and contrary winds, had given way to speedy steamships.

For Joseph Banks, his voyage round the world with Cook had been the high point of his life, and during his almost forty-two years as President of the Royal Society he loved to reminisce about it. Subsequent Presidents served briefer terms, and included Edward Sabine, Huxley and Joseph Hooker who had, like Banks, learned much of their science on their travels. Banks' journal was not published in his lifetime,[35] but many eminent scientific travellers[36] wrote up their reports in a readable way, giving rise to classics like Alexander von Humboldt's on South America, Meriwether Lewis and William Clark's on their journey across the USA, John Franklin's on arctic Canada, Darwin's on his voyage around the world, Henry Walter Bates' on the Amazon, David Livingstone's on central Africa, Alfred Russel Wallace's on the Amazon and on Indonesia, and Thomas Belt's on Nicaragua. The public appetite for scientific travel was huge, and the Royal Geographical Society, the US Congress which commissioned and published the *Pacific Railroad Reports*, and other bodies helped to feed it. Many such writings were not merely accurate descriptions of territory previously unknown to westerners, but also got across new scientific ideas.

Thus Humboldt showed that it was possible to map more than topography, starting physical geography with his isotherms and isobars; and also indicated visually how increasing altitude affects climate, so that mountain tops in Equador are like Spitzbergen. Darwin puzzled over the fossils of Argentina, and the fauna of the Galapagos Islands. Franklin's instruments, and those of Sabine, had to be modified to suit the extreme conditions. Bates noticed and described how butterflies from genera tasty to birds had evolved to look like distasteful species by natural selection, thus providing early examples of Darwinism in action.

Wallace in Indonesia not only independently hit upon the idea of natural selection, stimulating Darwin to publish in *The Origin of Species* (1859)[37] the mass of evidence he had been collecting over the years, but also saw how the animals and plants of different regions have different characteristics. Thus Bali and Lombok are separated by a narrow but very deep strait: Bali's fauna and flora are basically Asian, Lombok's Australian. Some travellers were like tourists on a cruise, based upon their ship and carrying a little bit of Europe and its assumptions with them. Wallace, living for long times among communities on the Amazon and in Malaysia, was (like Livingstone and some other scientific travellers) free of the casual racism so characteristic of the nineteenth century; his readers would have been challenged about their prejudices in regard to natives and savages, whom he did not regard as left far behind in an evolutionary struggle for existence. Readers of travel books could pick up important scientific ideas while following a ripping yarn of adventure and derring-do. John Herschel edited for the Admiralty a *Manual of Scientific Enquiry* (1849)[38] with advice from travellers (including Charles Darwin) on what to look for.

Stay-at-homes could also see some splendid pictures of exotic places. Illustration, based upon copperplates engraved or perhaps etched, had been very expensive before 1800, but the coming of wood engraving and then of lithography had made the printing of illustrations and maps much cheaper. Works of travel and natural history, and descriptions of experiments in chemistry and physics, were made much more attractive by a proliferation of illustration. But here, as with language, in works of science the pictures gradually came to look more like diagrams. The point was to get across things of scientific importance; and there was a tension between beauty and usefulness which was hard to resolve. The development of photography, a scientific art, downgraded topographical pictures; but for living creatures, where for scientific purposes a picture of a species of parrot rather than a portrait of Polly is required, an artist has continued to have an important place. We all enjoy good pictures, and they can be an important part of public understanding of, and relish for, science.

Science fiction was one kind of novel based upon science, but an interesting genre in the nineteenth century was the novel of religious doubt. Doubt was sometimes caused by scientific discoveries, though much more often and seriously (even probably in Darwin's own case) by bereavement, resentment at clerical pretensions, or unease generated by Biblical criticism. But the autobiographical novel by J.A. Froude, later a distinguished historian, *The Nemesis of Faith* (1848), caused a tremendous sensation; as later did Mary (Mrs Humphry) Ward's *Robert Elsmere* (1888), which was reviewed by Gladstone, and even given away in a soap promotion in the USA. Paley's *Natural Theology* was published towards the end of his life, in 1802. It was a great publishing success, with updated editions coming out regularly over half a century.[39] While it was not universally welcome, its utilitarian philosophy was more disliked than its general argument for Design, which was generally accepted.

The world closely examined seemed more and more to disclose the wisdom of God, which was also a title given to Jesus.[40] True natural theology should therefore show the harmony between natural and revealed religion, thus making science momentous and accessible to all in a religious age. Most popular science in the first half of the century was indeed permeated by natural theology, though the First Cause lying behind the law-governed world revealed by science might often seem very different from the Judeo-Christian loving Father. By 1900, popular science had become much more secular: agnosticism had become respectable, and people were even prepared to describe themselves as atheists, previously a term of abuse. For the first time, it was common to see religion and science in conflict. It is to natural theology and its decline that we now turn.

2 God's clockworld

The heavens are telling the glory of God, and the sublime science of astronomy discloses the wise and wonderful clockmaker.[1] This was something that most people took for granted, notably in George III's Britain; which meant that a little science would be good for everybody, and that popular science often took the form of religious apologetic. Natural history, the study of creatures great and small, was another source of wonder, and of faith: the microscope revealed the astonishing design and workmanship of the fly's eye, while navigators like Cook reported whales as huge as any imagined leviathan. Animals and plants, exquisitely adapted to their ways of life and environments, were in well-planned equilibrium. New discoveries made sense of old beliefs.

Clergymen of all nations and denominations shared in the culture of natural history. When the Spaniard Martinez Compañon was appointed Bishop of Trujillo in Peru in the 1780s, he had a handsome series of picture books prepared there to prepare him for his new abode. There are maps, and portraits of clerics and officers, laymen and women, Spaniards and Indians carrying on agriculture and trades. There are texts of songs, illustrations all in colour (and in various degrees of formality and naturalness) of plants and animals – including a frightful hairy snake, with a hungry head each end and an unwary bird caught around the neck by a monstrous crab. The artists differed in sophistication, but the result of their labour was to give the Bishop – and now us – a magnificent album illuminating the brave new world for which he was embarking.[2] God could be praised for the variety of creatures on the Earth, different in South America and in Spain despite similarities in climate; and also for consistency, for animals and plants everywhere were not wholly dissimilar – except for that alarming snake. The learned Bishop thus promoted or absorbed science as a part of high culture compatible with faith.

Spain and her colonies were Catholic territory; but while Great Britain was a Protestant island, there were especially in England numerous dissenters from the established Anglican Church. They had been tolerated since the late seventeenth century, but did not enjoy full civil rights; and in particular were barred from the ancient universities of Oxford and Cambridge. Down to the 1830s, these were the only ones in England, a poorly educated country compared to Scotland and even today with a continuing tradition suspicious of book-learning. The evangelistic campaigns of the Anglican John Wesley led ultimately to the formation of the Methodist Church as a separate denomination, and thus to a great growth in the number of dissenters. Right through our long nineteenth century, the tension between Church and Chapel, Anglicans and dissenters, was a crucial feature of English history. One of the few things upon which they could agree (apart from distaste for Catholics) was that the study of the creation could tell us something about its Creator: that all knowledge must tend to the glory of God. Wesley's preaching, with the message that all who repent can be saved, made religion a serious matter not only for Methodists but for everyone. It would be wrong to see most men of science using religion tongue-in-cheek to promote their work; they and others were believers, though not by any means always orthodox in their beliefs about God and nature.[3] Thus as well as the more-or-less orthodox, there were sceptics, doubting that any certainty was possible; and Deists, believing in a Creator or First Cause who did not intervene in the best of all possible worlds He had made.

Presbyterians and Congregationalists in eighteenth-century England and the American colonies were (like Wesley) chafing under the burden of strict Calvinism. Many rejected the idea of God's foreknowledge of our lives and fates, the doctrines of election and predestination that had been so important for the Puritans of the previous century. Instead, liberal-minded ministers and congregations moved towards Arianism or Socinianism, the view that Jesus was exceptional, but was not uniquely the Son of God: he was, as it were, adopted. They began to call themselves Unitarians, rejecting creeds, dogmas and doctrinal statements. As a group, Unitarians were liberal, promoting good causes, believing in progress and staunch supporters of education, especially in science, languages and other modern subjects rather than the traditional Latin and Greek classics. Notable among them was Joseph Priestley, one of the most radical ministers theologically, politically and in science, especially chemistry: he isolated and studied what we call oxygen.

Priestley believed that Unitarianism was the true and original Christianity, later corrupted by Greek philosophy; and that true religion and his chemistry both went with materialism. It was false to think that we were immortal souls confined within material bodies: rather, we were mortal, and at the end of the world God would by a miracle resurrect us bodily (as in medieval paintings and drama). In his *Disquisitions on Matter and Spirit*[4] (second edition, 1783) the atoms of matter were not inert billiard balls, but active centres of force, repulsive at short distances (to account for elasticity) and attractive at greater ones (to account for gravity). Priestley's interests in electricity and in gases thus complemented his religious beliefs. He had a stammer, a serious defect in a preacher; but he wrote beautifully and clearly and was a powerful advocate for Unitarianism, science and political liberty – notably full rights for dissenters.

His support for the French Revolution of 1789, as earlier for the Americans, was strident, and in the last 'Church and King' riots so far in Britain, his house in Birmingham was sacked by a mob on Bastille Day, 14 July 1791. He fled, but found himself unwelcome in London, where Banks,[5] as President of the Royal

Society, was doing his best to demonstrate that real science was not subversive[6] (as it seemed to have proved to be in France); and therefore emigrated reluctantly to the USA, ending his life under the Presidency of his friend Thomas Jefferson. Outside the Unitarian Church, in the political reaction and war beginning in the 1790s, Priestley's synthesis of religious belief and science did not catch on. For Banks and his associates in the elite scientific community based in London, science must go with respectable religion and due deference to the British constitution. Active, even thinking, matter would not do: another metaphor was needed.

Every bit as alarming was the dynamical idea that God had allowed the world to change and evolve without further interventions on His part. Jean Jacques Rousseau believed that savages were nobler than the calculating and double-dealing men and women in more civilised communities. If that were so, then orang-utans (and the very term in Malay means 'wild man of the woods') must be nobler still. The Scottish judge Lord Monboddo thought so; and Percy Bysshe Shelley's friend Thomas Love Peacock wrote a novel, *Melincourt*, in which the hero is an ape who becomes Sir Oran Haut-Ton, is very much the natural gentleman, though of superhuman strength, and is eventually bought a seat in Parliament – where his being able to vote but not speak makes him particularly valuable. Charles Linnaeus, the great Swedish classifier, had put the orang-utan in our genus (calling him *Homo sylvestris*) much nearer to us than apes are put today. Speculations about primitive peoples in remote places who still had tails were popular, as the youthful Thomas de Quincey tells us.[7] But it was in the writings of Erasmus Darwin that the notion of the evolution of mankind and of society was presented to a wide audience and became a part of popular science.[8]

Darwin was a very successful provincial doctor, who with Priestley, Josiah Wedgwood, James Watt and other men of science and manufacturers belonged to the Lunar Society of Birmingham. Meeting at the full moon, so that they would have light to get home afterwards, they discussed all kinds of scientific topics informally and without deference to religious authority.[9] Darwin took to poetry in *The Loves of the Plants*, popularising Linnaeus' botanical system in which the classification of flowers depended upon counting their sexual parts. Thus the Pentandria Digynia, which includes the gentians, has five male and two female organs in each flower – in bed together. In Darwin's robust eighteenth century, this could be the basis of a good deal of joking; and his poetry exploited these possibilities, and at the same time made the new botany familiar to a wide readership. Its lightness, optimism, vivid imagery, Deism and curious science made it attractive to both men and women. Indeed, by the 1790s, Erasmus Darwin was one of the most popular and widely read poets writing in English.[10]

The Loves of the Plants formed part of what became a bigger work, *The Botanic Garden* (1791) where the poetry was accompanied by an astonishing series of footnotes and endnotes, packed with curious information. For us, poetry that needs footnotes would be a turn-off; but didactic verse was clearly a genre popular with our ancestors. At the end of the century, the new poetic voices of William Wordsworth and S.T. Coleridge in *Lyrical Ballads*[11] (1798) created a furore and a new fashion. In 1803 Darwin's last poetic book, *The Temple of Nature, or the Origin of Society*, was published, a year after his death. By then Darwin's seemed one of the last voices of the Enlightenment, outdated, surviving into the epoch of the French wars, Romanticism and the evangelical revival spurred on by Wesley. Darwin proposed a progressively evolving world, and his text and notes are full of curious observations and arresting conclusions. There are even anticipations of natural selection in the struggle for existence: facing up to 'And one great Slaughter-house the warring world!', where even 'vegetable war' goes on endlessly as plants compete for soil and light. And yet things are improving and going upward. Darwin and his circle had little time for orthodox religion – the Wedgwoods[12] called their Unitarianism 'a featherbed to catch a falling Christian' – and there is no role for God to play as the world unfolds in conformity to the powers inherent in matter. With a war on, and a new intellectual climate, this would not do; and Darwin's poetry and broad sweep were mocked.

Robert Boyle in the mid-seventeenth century had been much impressed by the great clock at Strasbourg, which (after a number of rebuilds) still marks noon, local time, most strikingly with chimes, doors opening and shutting, and figures processing. To see why something happened it would be necessary to look inside and trace the mechanism. For Boyle and his contemporaries, the world was a great clock, and science a matter of finding its mechanisms.[13] This view gradually became popular, in what we call the Enlightenment,[14] at any rate among intellectuals. Taking something to bits and putting it together again, analysis and synthesis, became the ideal in chemistry and physics, even if not fully practicable in zoology.

Clocks are driven by a mainspring or a weight, and while it might be alright to see God as the mainspring of the world, in terms of terrestrial politics such a view went with despotism. William Harvey's comparison of the King to the heart probably delighted his autocratic patron Charles I; and Louis XIV might see himself as the driving force behind a kingdom running like clockwork. But, in Britain and America, by 1776 a rhetoric of checks and balances came to replace the more complex imagery of clockwork in political discussion. This was however just the time when the chronometer had been perfected for the discovery of longitude. Local time was compared with that shown by the chronometer set on the meridian at Greenwich, and the difference gave the longitude – one hour corresponding to 15°. The success of John Harrison, and then other makers, in fashioning clocks which kept good time in a pitching and tossing ship, voyaging for weeks or months through extremes of temperature, caught the public imagination and made the mapping of Cook and his successors very much easier.[15]

Clockwork thus became fashionable again in time for William Paley to write his celebrated *Natural Theology*, first published in 1802.[16] This book was the culmination of a life devoted to writings defending Anglican Christianity. He imagined finding a watch on the path, picking it up and noting how well all the parts work together. There is glass so that we can see the hands; when we open the case, we see intricate brass and steel work, nothing redundant and ingenuity everywhere. How absurd it would be to say that atoms had come together by chance to generate a watch. There could be no doubt that it had been made; it displayed design and craftsmanship. The rest of the book is a series of arguments, cumulative rather than rigorous, to show that the world is an enormous clock, and that the animals and we are little watches; and that God has chosen to make the world the happiest possible. We enjoy eating, for example, which might be merely tedious refuelling; and even the shrimps seemed to Paley to enjoy their swim in the then-unpolluted waters off the coast of Cumbria. From all the examples of contrivances that Paley gives, his readers would absorb a good deal of the biology, anatomy and physiology of the day.

They would also have learned some non-mathematical astronomy, for Isaac Newton had demonstrated the power and wisdom of God that lay behind the planetary orbits and the simple laws to which they were subject. Paley was, like Newton, a Cambridge man, and his well-written and accessible books on moral and political philosophy and evidences for Christianity became required reading there. William Pitt, the Prime Minister, admiring his *Moral Philosophy*, called him 'the best writer in the English language';[17] Charles Darwin found *Evidences* and *Natural Theology* the only useful and congenial part of his formal courses in Cambridge. At the new University of Durham, founded in 1832, the same rule applied. Some editions of *Natural Theology* have questions at the back to help students swotting for examinations. Paley, who had a career in the Anglican Church taking him from Cambridge to Carlisle, and ending up with posts in the industrial city of Sunderland and in Lincoln, was careful to note that natural religion based upon science could only be a preparation for revealed religion, in the Bible, and not a substitute for it. In France Robespierre had sought to replace the feasts of the Church with a Festival of the Supreme Being; that is, to establish natural religion. Paley would have none of that.

A problem was the evil and pain in the world, for Paley could not fail to describe the contrivances by which animals snare and eat each other. His solution was to adopt utilitarianism from the otherwise heretical Priestley, who had popularised the phrase 'the greatest happiness of the greatest number'. On balance, the world was one in which God had maximised pleasure and minimised pain, as we should seek to do in our moral lives. Carnivores were for Paley machines for euthanasia: the antelope in late middle age was spared the pains of arthritis and decay by being gobbled up by the wolf – it was all over quite quickly – and the wolf had the pleasure of a good dinner. A world with carnivores was therefore happier than one without. Ichneuman flies, whose larvae slowly devour living caterpillars within which they live, and gadflies which lay their eggs beneath the skin of cattle, were a little more of a problem (they darkened Darwin's poem); but nobody could deny the ingenuity of the Creator, even if He moved in mysterious ways. As consolation to the invalid, the widow or the orphan whose parents were eaten by wolves, the message that on balance the world is the happiest possible is never very effective; but Paley, like

Priestley, wrote very well, and with real enthusiasm. Generations of students and other readers absorbed much science from him; and the message conveyed was reassuringly that of Francis Bacon, that real science rightly understood must support true faith and sound political institutions.

War between France and Britain broke out in 1793, and lasted with only a brief half-time break in 1802 (the Peace of Amiens) and another in 1814 until the battle of Waterloo in 1815. A whole generation grew up who had known nothing but world war, for there were battles in the West Indies, the USA, Egypt and India as well as in Europe, from Portugal to Moscow, and Britain took over the Dutch colonies in Indonesia and at the Cape of Good Hope. Napoleon, who seized power in November 1799, brought the Pope to France where he was kept under what was in effect house arrest at Fontainbleau. Napoleon signed a Concordat, restoring Roman Catholic worship in France; but to outsiders, his Empire looked as ruthless and irreligious as the revolutionary governments had been. Science, in the writings of Denis Diderot, Jean d'Alembert, Voltaire and Rousseau, seemed to have been the corrosive agent that had undermined the ancien regime; and in revolutionary and Napoleonic France, it duly flourished mightily.

Indeed, Paris was the world's centre of excellence in science right through the years of upset and war. Science there could be communicated as modern and republican. In mathematics, astronomy, experimental physics, chemistry, medical sciences, zoology and botany, the French were the leaders. The Academy of Sciences was closed, and the great chemist and fat-cat taxman, Antoine Lavoisier, executed in 1794, under the Terror; but scientific organization was soon revived in the First Class of the new Institut.[18] Men of science rallied to the republic, supervising the melting of church bells into cannon and other preparations for total war; while Napoleon fancied himself as a scientific man, and was duly elected to the Institut. There were bright individuals in Britain and Germany, but outside medicine there were few paid posts in science and nothing in the educational system to match the new and meritocratic École Polytechnique which combined teaching with research, and trained engineers and men of science.

It was the new industrial economy of Britain (in places like Birmingham and Sunderland) that defeated the French; for, while France led in science, Britain had the new technology of steam engines and textile machinery, based on organised common sense rather than recondite research or the latest theory. Dissenters were prominent in industry, and in banking: capitalism was underway. Bacon had believed that science would enable mankind to evade the curse put upon Adam and Eve when they were expelled from Eden: it would reduce labour, pain and disease, and make the world fruitful. Science as a useful activity that would improve standards of living had therefore a religious aspect in Regency Britain.

In 1802 the young Humphry Davy gave his inaugural lecture at the Royal Institution, which had been founded in 1799. It was a splendid performance, presenting a sexy picture of men of science penetrating to the bosom of the Earth, and the bottom of the sea, to allay the restlessness of their desires. The audience was delighted, and Davy's lectures became a feature of the London scene; Albemarle Street had to be made one-way on evenings when he was holding forth, to prevent gridlock. What they must have taken away from this lecture was a new gospel of applied science, with Davy as its apostle.[19] The many problems facing Britain, rapidly industrialising, at war, often hungry and always unequal, could all be solved in time by the progress of science and its application to practical questions. The refined common sense which was science meant more than the inspired rule of thumb followed by the early engineers; it involved systematic research in the laboratory. The future lay with those who could harness science to agriculture and industry; and the Royal Institution, with its wealthy members attending lectures (with their wives and daughters, for science was a part of high culture too) and the research laboratory in the basement, embodied this vision. There in 1815, following a terrible explosion in a pit near Sunderland, Davy (Britain's leading chemist, already given a prize by the Parisian Institut, despite the war) invented the safety lamp for coal miners which made him a world-famous philanthropist.

Allusions to the wisdom of God, and the need for us to do experimental science rather than pore over the Scriptures if we wanted to understand and master the physical world, were a part of Davy's rhetoric. As an associate of Coleridge and Wordsworth (he oversaw the printing of the second edition of *Lyrical Ballads*), he delighted in mountain scenery. He was also a keen fisherman, and much more likely on Sundays to be worshipping God by the banks of a trout stream than in church. Schooled by Coleridge, he had had a mystical experience of nature when about twenty-one (1800), and he later wrote a rhapsody:

> Oh, most magnificent and noble nature!
> Have I not worshipped thee with such a love
> As never mortal man before displayed?
> Adored thee in thy majesty of visible creation,
> And searched into thy hidden and mysterious ways
> As Poet, Philosopher, as Sage?

Romantic pantheism, worship of sublime and beautiful Nature rather than of a transcendent God, was a feature of Davy's writings;[20] and also of later men of science, such as John Tyndall, heirs all of them of both the Enlightenment and the Romantic Movement. Unlike Davy, many later pantheists were to be contemptuous of organised religion; but, in his time and later, a heightened awareness of beauty and awe, conveyed in writings, lectures and pictures, was an important spur to science.

Davy's father and uncle had died of heart attacks in their forties, and he, in his turn, had a stroke in 1826 when he was just forty-eight, at what should have been the height of his career, as President of the Royal Society. Sent abroad to the warmer climate of Italy, and spending the summers among his beloved

Julian Alps in what was then the Austrian province of Carniola (now Slovenia), he began composing dialogues: first on fishing, and then when this was published and successful, on life and death, knowledge and wisdom. He died after another stroke in 1829, bequeathing these dialogues to posterity and making his devoted and hero-worshipping younger brother his literary executor. They were published as *Consolations in Travel, or the Last Days of a Philosopher* in 1830; and described by Georges Cuvier, who wrote Davy's obituary as permanent secretary of the Parisian Institut, as the work of a dying Plato.

Religion pervades the volume. It begins with a vision of progress, a dream in the Colosseum; and continues with discussions of reincarnation, birth and death, time and change, and science as a career. Some aspects of some of the characters fit with friends and colleagues of Davy's, but basically they are all aspects of his personality and do not come to life as separate individuals. In that respect, the book cannot be called a literary success; but it succeeds in putting across a scientific world-view characteristic of its time, and it went on selling in editions, and in new translations, for many years. Later editions were embellished with engravings, from drawings done by the wife of the eminent geologist, Sir Roderick Murchison – persuaded to take up science (as compatible with field sports) by Davy.[21] The religion is not orthodox – it involves the planets as places on which souls are reincarnated – but it is a factor in making the science momentous, and it allowed Davy to believe that, when his soul left his worn-out body, it would move into a higher sphere. He came out firmly against materialistic physiology – plausible in the dissecting-room, but soon dismissed with a walk in green fields – and evolution in favour of a progressive interpretation of the geological record and presenting the chemist as a god-like figure creating new substances for the benefit of mankind. This was popular science of a high order, setting out to be real 'natural philosophy', making sense of the world, using the high faculty of reason rather than mere mechanical understanding, or the fancy associated with mere hypotheses.

While Davy was dictating his dialogues to the bored medical student, John Tobin, who was his companion in these last travels,[22] the Earl of Bridgewater died in February 1829. He had been heir to a canal fortune, and he bequeathed the large sum of £8,000 in trust to the President of the Royal Society to nominate authors to write books on the power, wisdom and goodness of God as manifested in the Creation. Davies Gilbert (formerly Giddy), the Cornish MP and mathematician who had succeeded Davy, discussed the bequest with the Archbishop of Canterbury and the Bishop of London. They settled upon finding eight authors, who would each take a different branch of science. This was to be a more orthodox exercise than Davy's *Consolations*; most of the authors were Anglicans, but Thomas Chalmers was a Presbyterian and a leading light in the Church of Scotland, and later in the Free Church. He had established a great reputation with lectures on the Newtonian universe, and God's wisdom and goodness displayed therein, in Glasgow. His *Treatise* was concerned with the adaptation of external nature to our moral and intellectual constitution. John Kidd, Professor of Medicine at Oxford, wrote about our physical condition;

William Whewell of Cambridge about astronomy; Charles Bell, from Edinburgh, on the human hand; Peter Mark Roget (now famous for his *Thesaurus*) on physiology; William Buckland on geology; William Kirby on habits and instincts, notably of invertebrates; and William Prout on chemistry, meteorology and digestion – a rag-bag of sciences left over, where he had interesting things to say about the atoms of matter. The volumes came out over several years, as their authors finished them, and thus not exactly in order; and they proved (surprisingly to publishers who had turned the project down[23]) to be a publishing success story, selling very well for many years, first in expensive editions, and then in cheaper reprints.

The whole series, appearing between 1833 and 1836 and written by prominent members of the scientific establishment, demonstrated the continuing power of natural theology as a splendid vehicle for popular science at a time when, in Oxford (and to a smaller extent in Cambridge), there was a great revival of religion in a form largely indifferent to science. The Reform Bill of 1832, which extended the vote to the middle classes, was the culmination of a process that had in the 1820s brought full civil rights to Protestant dissenters and then to Roman Catholics. It seemed to John Keble, poet and Anglican clergyman, that this represented national apostasy: and he said so in an Assize Sermon preached before the judges in Oxford in 1833. He was soon joined by E.B. Pusey and John Henry Newman in a campaign to restore the fortunes of the Church of England, and to emphasise tradition, liturgy and the calling of the priesthood – the Catholic rather than the Protestant aspects of its history.[24] Liberal-minded and scientific clergymen, like Buckland who was Reader in Geology, found themselves sidelined, as for a decade, undergraduate students and their teachers enthusiastically took sides in religious controversy, great numbers flocking especially to Newman's sermons until, in 1845, he joined the Roman Catholic Church.

In the world outside, the most interesting Bridgewater Treatises were probably Whewell's and Buckland's. Astronomy had been dominated by the French, and the greatest figure had been P.S. Laplace who, when asked by Napoleon what role there was for God in his system, replied that he had no need of that hypothesis. Whewell had to restate the view of Newton: that using astronomy to demonstrate the wisdom of God was reasonable, the solar system and the stars showing every sign of design, forethought and contrivance just like a clock. Immanuel Kant had argued that while the starry heavens and our conscience impelled us towards belief in God, we could never prove His existence. All the so-called proofs failed when closely and coldly examined. The Dutch mathematician and astronomer Christiaan Huygens had earlier commented on religious apologetic, 'How far this stands from the power of persuasion afforded by mathematical proofs!'[25] While, however, Kant's work undermined tight deductive arguments like those used by Rene Descartes, it did not really affect cumulative arguments like Paley's, which are not like a chain which breaks at its weakest link, but like a rope (as are political manifestos, or legal pleas). Particular cases of apparent design may or may not convince a reader; but, like a cord with some fraying strands, the argument will still bear weight. Most Bridgewater authors followed Paley's strategy.

Whewell[26] was knowledgeable about philosophy (and everything else), and therefore put his argument in a different form. If we believe in God, then we can find out more about Him both by reading and pondering the Bible, and also by studying the world He has made. The leap of faith, while not unreasonable, is not a matter of logic; but after we have made that leap in the dark, logic must inform the argument from up-to-date science to knowledge about the Creator. In analysing scientific reasoning, Whewell believed that a similar process went on: it was crucial to get the right end of the stick, the appropriate basic idea, intuitively; and then it could be refined and corrected by experiment and mathematics. Purely deductive reasoning, the pure mathematics which Laplace and others developed to handle astronomical calculations, led to arrogance: or so Whewell argued in his book, and also in Cambridge where he became a great man (Professor of Mineralogy, and then of Moral Theology, and finally Master of Trinity College) and upheld the position of applied (or 'mixed') mathematics, with its empirical connections, in the syllabus.

Geology, with its vistas of deep time,[27] also attracted the attention, and sometimes raised the hackles, of the religious-minded in a way that chemistry, mineralogy and digestion probably did not. Buckland, like Whewell, was an ordained clergyman of the Church of England.[28] This was the usual step for anybody following an academic career, or one in the newly developing Public Schools; it brought moral and intellectual authority, but entailed defending church doctrine. Buckland had in his *Reliquiae Diluvianae* (1823) inferred from the bones of hyenas found in a cave in Yorkshire that the cave had been their den, and that they had been drowned in Noah's Flood. They were a different species from modern African hyenas, but very similar; and in particular, they crunched the bones of their prey in just the same way, and their whitish faeces were alike – Buckland carefully observed hyenas in a menagerie. Cuvier had reconstructed extinct creatures from fossil bones found in the quarries of Montmartre when Napoleon was rebuilding Paris in Imperial splendour, and had concluded that there had been a succession of faunas and floras in France, separated by catastrophes. To Buckland, the first geologist awarded the Copley Medal of the Royal Society (its highest honour) for his work, the Flood was the latest of such cataclysms. The medal was presented by Davy, who remarked that this was the first time in its ninety-year history that it had gone to a geologist: the science was prestigious and popular, and its connections with *Genesis* made it exciting for everybody.

Nobody at Oxford then could study for a degree in geology (or indeed anything except for Classics and Philosophy – 'Greats' – or in some cases Mathematics), but Buckland and other Professors gave lectures to which anybody interested could come. Among those who did was Charles Lyell, intended for the law but seduced by geology. He believed that his professor (and Cuvier) had been misled into perceiving series of catastrophes because they had not allowed long enough for the ordinary processes of uplift, deposition and erosion to do their work: they were prodigal of violence because parsimonious of time. Lyell's *Principles of Geology*, beautifully argued with a lawyer's skill in presenting a case, appeared in 1830–3 and proposed that past changes should be explained exclusively in terms of processes now operating.[29] This meant a history of hundreds of millions of years, in apparent straightforward conflict with the chronology of the Bible: where adding up the ages of the various patriarchs when they begat descendents led to a date for the Creation of 4004 BC, accepted by the literal minded, though not by Buckland.

Buckland changed his mind, in accord with his pupil's reasoning: Lyell remarked[30] how scientists happily were not expected like clergy 'to retain for ever the same views'. In his *Bridgewater Treatise*[31] Buckland, accepting much of Lyell's gradualism and long time-scale, argued (with the support of Pusey) that the first verse of *Genesis* is a theological statement: In the beginning, God created the Heavens and the Earth. The next verse describes how the present state of things came about: mastodons and dinosaurs were not just pre-historic, but pre-biblical; millions of years, as shown in the rocks, had passed while the Earth was reaching the state appropriate for mankind – it was cooling down, and deposits of iron ore and coal were being laid down. Buckland was very interested in dinosaurs, their footprints, and even their bowel movements (following up his studies of hyenas), making deductions from their fossilised faeces. They were not poor designs, mistakes made by a Creator blundering towards mankind by trial and error, and like a doctor burying His mistakes: on the contrary, they were apt for their time and place, like the animals alive today. Through the geological record, for Buckland though not for Lyell, we do see progress, design and forethought. Properly understood, geology was a great prop for Christianity.

Buckland's book was in two volumes, the second being plates on which Buckland was said to have spent his £1,000; the coloured frontispiece folds out, and is over a metre long. It shows the immensely lengthy geological epochs, and their characteristic fossils. Not everybody liked his exposition of *Genesis*; some favoured instead the idea that the 'days' of creation were a thousand ages each; but, on the whole, the book was a great success, making geology momentous and also reassuring. Buckland was soon afterwards a convert to the idea of the Swiss geologist Louis Agassiz that there had been an Ice Age. Alpine glaciers had begun retreating at the end of the eighteenth century, and their characteristic moraines and polished rocks became clearly visible – and exactly similar phenomena could be seen all over northern Europe. Agassiz's wonderful plates of glaciers, with explanatory transparent overlays, made his book powerful reading.[32] Reading popular works on geology got people looking harder at scenery, and was recommended for landscape artists. There clearly had been catastrophic changes of climate, even though they had not happened in the twinkling of an eye. But Lyell, studying the fossils of geologically-recent Italy, realised that between epochs there was no complete cut-off: like centuries or dynasties in history, the Eocene or Miocene periods were convenient divisions rather than sharp boundaries where everything changed.

Buckland's paper for the Royal Society was embellished with copperplate engravings; but when he was preparing his books he turned to the new and much cheaper method of illustration, lithography. On a suitable slab of stone, the artist makes the drawing with a wax crayon. The stone is then wetted, and inked with a greasy ink – which sticks only to the waxy parts, so that the drawing can be printed. This meant a revolution in the appearance of scientific books, notably in geology, because they became much better illustrated, and thus clearer and more attractive. Not only that, but the scenes from deep time that they contained brought home to everyone the primeval world of dinosaurs, the dragons of the prime which tare each other in their slime, and other extinct creatures.[33] Moreover, these pictures seem to have owed a good deal to biblical illustration: artists like John Martin, who had painted the destruction of Sodom and Gomorrah, and the writing on the wall at Balshazzar's feast, also depicted the country of the iguanadon. The garden of Eden and the story of Noah also provided a tradition of portraying exotic plants and animals, at peace or in terror. Pictures brought extinct creatures to life, and launched our continuing obsession with dinosaurs: large models of which then became a feature of the Crystal Palace when, after the Great Exhibition of 1851 in Hyde Park, it was re-erected at Sydenham. There, some leading geologists dined inside the iguanadon (before its back was put in place).

The tradition of popular science in the form of natural theology continued right through the nineteenth century. One of its greatest exponents was the Scottish stonemason turned geologist, Hugh Miller, whose vivid writings created enormous interest in fossils and minerals.[34] He was especially eloquent about the often armoured and extraordinary-looking fish that are abundant in the 'Old Red Sandstone' of Scotland, writing a whole book on the Asterolepis of Stromness (in the Orkney Islands), under the title *Footprints of the Creator* (1849). For the American edition the following year, Agassiz wrote an enthusiastic preface. With Chalmers, Miller was one of the leaders of the Free Kirk in the Great Disruption of the Church of Scotland, when in 1843 the more evangelical and independent-minded broke away from the rump over the question of the appointment of ministers by congregations rather than lairds.

His *Footprints* was directed against the anonymous *Vestiges of the Natural History of Creation* (1844),[35] an evolutionary work which to Miller, Sedgwick and most clergy and men of science, seemed atheistical, materialistic and riddled with error. It created a sensation. It was written by Robert Chambers, who with his brother William was a prominent Edinburgh publisher, taking advantage of the new markets for books opened up by education (the March of Mind), cheap paper, steam printing and case bindings – which brought hardback books into being, and reduced the opportunities for craft bookbinders like the young Faraday, who had bound to the customer's specification, in leather or buckram, books which were expensive and were sold in sheets or in boards, like a paperback or a modern learned journal.

Chambers read the books he was publishing and selling to individuals, as well as to Literary and Philosophical Societies, Mechanics' Institutes, and other libraries. He was very careful to keep the authorship of *Vestiges* a secret, and he found through an intermediary another publisher for it. It was a story of cosmic evolution. He believed that the spiral nebulae just perceived through the latest telescopes were solar systems in the making; that from a mass of fire mist, which began to swirl, the Sun and planets had been formed. As the Earth cooled, a living filament or globule was generated, rather as a creepy-crawly was just reported to have been spontaneously generated in an electric battery, and progressive evolution then in due course led to humans. Embryology was, with the development of microscopes, a new and exciting science: the mammalian ovum had only recently been observed for the first time, by Karl von Baer.[36] Chambers believed that evolution happened jerkily when an embryo stayed longer on the main line of development (this was the period of railway boom) rather than going off on the branch that its parents had taken. Thus from some ducks' eggs, (several) platypuses had in due time emerged; and from their eggs, rats. Australia still had its curious fauna, Chambers believed, because it was a young country.

He finished with reflections on mankind, shown by the development of statistics to be much more predictable in the mass than anybody had supposed: determinism prevailed throughout the world. The book was not strictly-speaking atheistic; Chambers' God was the First Cause, the creator who had started and programmed the whole thing. But He was not the loving father of the Bible. Chambers therefore managed to offend everyone. Astronomers were abandoning the nebular hypothesis; physiologists did not believe the just-so story about ducks, platypuses and rats; geologists were finding that the rocks of Australia were very ancient; while moralists and clergymen were appalled at his picture of humans as predictable, their free choices an illusion. The book caused a immense excitement,[37] and the crop of hostile notices helped to sell it. Public understanding of science, as of religion, has always been forwarded by heretics and sceptics.

The effect of *Vestiges* was to promote the older natural theology; and to make respectable men of science focus upon detail rather than the big picture. But, especially in London, there were by the 1840s those happy to reject these conservative ideas, supportive of the status quo, and to fall in with the world-view of *Vestiges*. Once again, science might be subversive. The 'hungry forties' were a time of economic depression and political unrest all over Europe, culminating in the revolutions of 1848, which affected almost every country – even Britain had big Chartist demonstrations, which frightened the authorities. Just as those popularising science do not always pick upon those aspects which leading scientists wish they would, so when it was – and is – done in terms of religion, the science may be unorthodox. There is no doubt that it was the mix of religion and science in *Vestiges* that made it the great success it was, though it brought opprobrium upon its unknown author – various people, including Prince Albert, were supposed to have been responsible for it. To link scientific knowledge with values, revolutionary or traditional, is a surefire way of attracting attention to it. The evolutionary world did not run like clockwork: it was open-ended, with unpredictable outcomes, and hence attractive to radicals confronting a sclerotic establishment in politics and medicine.[38]

Until the middle of the nineteenth century, however, it was usually the traditional values and sufficiently orthodox religious views that were most prominent. Thus Thomas Dick,[39] a Scottish schoolmaster alarmed at how unimportant religion seemed to be in Mechanics' Institutes, wrote *The Christian Philosopher* in 1823 and *Celestial Scenery* in 1839; these were very effective works in their day, popularising science in a palatable religious form. They were enormously influential in Britain and the USA; young David Livingstone was a convert to Dick's views. Somewhat similarly, in the more technical *Chemical Catechism*[40] of Samuel Parkes, there are copious footnotes some of which are suffused with Unitarian piety; though begun for his daughter, it was written to encourage parents in the expanding middle class to get their sons to learn chemistry and enter industry – religion and wonder help to make the pill of instruction go down, and promise usefulness.

We are prone to think of evangelicals, the most important group within society in the early nineteenth century, as biblical literalists dogmatically opposed to scientific theory in the way that their modern successors, the 'religious right', promote 'creationism'.[41] But nineteenth-century evangelicals were mostly not like that, although there were some 'scriptural geologists' who were appalled by what they saw as Buckland's playing fast and loose with *Genesis*. But, to an evangelical, 'natural theology' was all too likely to lead to mere Deism; while real religion meant salvation from sin and acceptance by God the Father through atonement by Jesus on the cross. The particular character of evangelical understanding and popularising of science has been studied by Aileen Fyfe. Accepting that the study of God's works was very properly to be undertaken in conjunction with the study of His word, evangelicals promoted what has been called 'theology of nature'; and the Religious Tract Society published many works that were Christian in 'tone', making explicit their commitment to orthodoxy in a way that authors in other series, like the Chambers' (and even those published by the Society for Promoting Christian Knowledge) did not.[42] They were usually anonymous, the publishing house rather than the author being important. Such books were written in defence of astronomy for believers, not in defence of religion for astronomers.

Parkes' science of chemistry, aiming at improving the world, was always problematic for the theologically-minded.[43] Were synthetic materials the same as natural ones, or did they just appear so but differ in essence? Might life itself be synthesised?[44] These issues, real to us, go back to medieval debates on alchemy but were revived by Mary Shelley in her *Frankenstein* (1815), and Davy in his remarks on the chemist as creative. Then, in 1838, Hannah Acton, in memory of her husband Samuel, endowed a prize for an essay on the wisdom and goodness of God, to be awarded every seven years by the trustees of the Royal Institution. The first winner, in 1844, on the subject of Chemistry, was George Fownes. His essay was published by John Churchill, the medical publisher who brought out *Vestiges* in the same year; Fownes was indeed to help in revising subsequent editions of that book. In his own book, Fownes invited readers to look for the simple and harmonious laws of chemistry in common and everyday processes;

declared in Puritan vein that labour is in fact a blessing and not a curse, preserving us from idleness; and concluded with the reflection that the real attraction of science is the contemplation of truth for its own sake:[45] 'Nowhere throughout the whole creation is the goodness of the Almighty more conspicuous than in the means he has provided for revealing himself to his intelligent creatures, by conferring upon them these very powers of discovering truth, and appreciating its beauty and loveliness.' Those who thought that chemistry was 'stinks' would be surprised; but the reader would in fact have got a good picture of the state of the science at the time. It was an effective popularising work, bringing before its readers the new organic chemistry especially associated with Justus Liebig.[46]

Liebig's own *Familiar Letters on Chemistry*[47] were rather different, being written by a most eminent practitioner, but still suffused with what is to our eyes a surprising concern for an overview in which spiritual as well as material concerns are prominent. Liebig had no time for the *Naturphilosophie* of his older contemporaries, Friedrich Schelling[48] and Georg W.F. Hegel,[49] seeing it as an intellectual Black Death that was diverting people from genuine empirical science into empty philosophising. For Hans Christian Oersted, however, it provided the world-view he needed. If everywhere there were polar forces, and the dialectical clash of apparent opposites generated new syntheses, then electricity and magnetism must be connected.[50] To the surprise of contemporaries he demonstrated the link, making a compass needle twitch as he turned the current on and off in an adjacent coil of wire. His reputation made, he wrote essays collected and published as *The Soul in Nature*[51] in which he sagely propounded a liberal spiritual view of the world: to Charles Darwin, firmly in a more empirical tradition and admiring Paley, this book was 'dreadful';[52] its English translation represents the last twitchings of romantic science.[53]

Much more sober was the *Religio Chemici* of George Wilson,[54] professor of technology in the University of Edinburgh (where he had to devote much of his time in explaining what the word meant).[55] An invalid, he died young, and the book was assembled and published posthumously; the title echoes the *Religio Medici* of the seventeenth-century physician Sir Thomas Browne,[56] but the book is only partly about religion, and includes valuable biographical essays. Wilson's life included much suffering, and he could not write with the easy optimism of many natural theologians (or theologians of nature) who, like Paley, had enjoyed robust health and a steady income. He saw the dark side of life, as did Charles Darwin, following the death of his daughter and was well-aware of the competition and predation inherent in the natural selection he perceived going on.

We have been led to expect resistance by clergy, and religious believers, to his evolutionary ideas, but Symonds' *Old Bones*[57] shows that this was by no means always the case. Based upon talks given by the distinguished parson–naturalist, it indicates how he at least admired Darwin and his work, and had no great problem in incorporating his ideas into his Christian framework (like most church people today), and making an attractive little book out of fossils, their history and their relationships. There were people who went to hear Darwin's disciple Thomas Henry Huxley and Richard Owen,[58] responsible for the Natural

History Museum in South Kensington, angrily debate evolution; but Symonds chose to emphasise their wide measures of agreement.

Owen had taken up from Germany a biology of 'types', in which actual animals could be seen as different realisations of ideal archetypes, and evolution as the process whereby different types died out and came into being.[59] He instigated the translation of Lorenz Oken's *Physiophilosophy*, a classic of Naturphilosophie.[60] Natural selection was for him a crude and materialistic version of such a belief. Paley's clockwork universe was a casualty of the Romantic Movement. It had been replaced by an evolving world, of Heracleitean flux and dynamic equilibria rather than fixity and mechanism. But Owen, like many of his contemporaries, notably in Germany,[61] saw no great problem about incorporating these newer understandings into a framework where natural theology was still alive. It was still, after all, teleological in spirit. Tennyson had indeed played with the idea of types, in his great poem *In Memoriam* where science yields an alarmingly glum world-picture, threatening humane understanding:[62]

> Are God and Nature then at strife,
>   That Nature lends such evil dreams?
>   So careful of the type she seems,
> So careless of the single life . . .
>
> 'So careful of the type?' but no.
>   From scarped cliff and quarried stone
>   She cries, 'A thousand types are gone:
> I care for nothing, all shall go.

The idea of the 'type' was deeply entrenched in theology, where characters and events in the Old Testament were perceived as foreshadowing those in the New. For Henry Drummond, preacher and evolutionist, whose *Ascent of Man* gave an optimistic tone to Darwinian evolution, types could reunite scientific and religious understanding.[63] Philip Gosse had been out on a limb in his famous *Omphalos* of 1857, suggesting that just as the motherless Adam and Eve would have had navels, and the freshly-created trees in the Garden of Eden rings, so the Earth, created in 4004 BC, had fossils falsely suggesting a long past.[64] Natural theology undoubtedly weakened in the second half of the nineteenth century, as the confidence and professional standing of scientists increased and clergy were forced onto the defensive; but as a way of propagating, if no longer generating, scientific understanding, it was still important.

Oersted's work was taken up by Michael Faraday, a man of deep personal faith who generally (in the manner of his small Sandemanian sect) kept his beliefs to himself.[65] He did his research in electrochemistry and then electromagnetism, making our world possible, at the Royal Institution in London, founded in 1799 to popularise science at a high level; and he lectured there, to adults and then very famously to children at Christmas times. Huxley also lectured there; and it is to these and other lecturers, getting science across entertainingly, that we now turn.

3   Holding forth

Scientific lecturing, to the nobility and gentry and then increasingly to the middle classes in public buildings, was a feature of eighteenth-century Britain, in a new world of leisure, civility and prosperity. It was a way of earning a living. In universities, the sciences occupied a firm place only in the medical faculty. By this time, astrology was no longer believed to be significant in diagnosis and healing, and so there was no scope for astronomers to benefit; but chemists, botanists and zoologists could make a career teaching medical students. One result was that such eminent men of science as William Hyde Wollaston the crystallographer and metallurgist, Joseph Hooker, the Director of Kew Gardens and 'Darwin's bulldog', Thomas Henry Huxley, were all doctors by training, who supported themselves at times in their lives by practising medicine. None of them became Professors of Medicine, but Linnaeus in Uppsala and Joseph Black the chemist in Edinburgh had done that. Formal lectures to students have to cover a syllabus; and in Edinburgh and elsewhere the professor's salary was heavily dependant on the fees he received from the students who attended. This meant that those professors who covered the most essential parts of the medical curriculum, and who were lively speakers, got a good income. For promoting their careers, in what was primarily a clinical discipline, their research was of little or no significance. They got on as teachers, though the lectures were all too often 'dull and humdrum'.[1]

The nobility and gentry, as well as artists, used to attend the dissections carried out in the anatomy theatres of Bologna, Padua and Uppsala up to the eighteenth century. Even later, in a generally more queasy age, medical experiments, especially if extra-curricular, could be riveting to audiences. Thus we read about demonstration experiments performed in 1823 in the anatomy theatre of Glasgow University by Dr Andrew Ure, of the nearby Andersonian Institution, author of a chemical dictionary and a promoter of applied science generally:[2]

> The subject of these experiments was a middle-sized, athletic and extremely muscular man, about thirty years of age. He was suspended from the gallows nearly an hour, and made no convulsive struggle after he dropped; while a thief, executed along with him, was violently agitated for a

considerable time. He was brought to the anatomical theatre of our university in about ten minutes after he was cut down.... A large incision was made into the nape of the neck, close below the occiput.... A profuse flow of liquid blood gushed from the wound, inundating the floor ... the pointed rod connected with the battery was now placed in contact with the spinal marrow while the other rod was applied to the sciatic nerve. Every muscle of the body was immediately agitated with convulsive movements resembling a violent shuddering from cold.... On moving the second rod from the hip to the heel, the knee being previously bent, the leg was thrown out with such violence as nearly to overturn one of the assistants. [Connecting differently] Full, nay, labourious breathing instantly commenced. The chest heaved and fell; the belly protruded, and again collapsed, with the relaxing and retiring diaphragm. In the judgement of many scientific gentlemen, this respiratory experiment was perhaps the most striking ever made with a philosophical apparatus. Let it also be remembered that for full half an hour before this period, the body had been well-nigh drained of its blood. [Reconnecting again] Every muscle in his countenance was simultaneously thrown into fearful action; rage, horror, despair, anguish, and ghastly smiles, united their hideous expression in the murderer's face: surpassing the wildest representations of a Fuseli or a Kean. At this period several of the spectators were forced to leave the apartment from terror or sickness, and one gentleman fainted.

We are in the gothick world of horror stories, and Mary Shelley's recently-published *Frankenstein*. But even surgery for medical students, like the young Charles Darwin soon to go up to Edinburgh,[3] where there was also grave-robbing for anatomy specimens (short-circuited by Burke and Hare, who murdered in order to sell the bodies to the medical school), made serious medical training a grim business.[4]

Oxford and Cambridge had science professors too, but they were not closely linked to medicine (which was not actively taught there anyway). Richard Watson, who later became a Bishop, took up the chair of chemistry at Cambridge in 1764.[5] He had admitted before his election that he knew nothing about the science, and had never even seen a chemical experiment; but he was a bright man, a Fellow of Trinity College, and somewhat bored with mathematics. He duly set to work, studying lead, zinc and gunpowder, heat and solutions, and the distillation of coal. Acquiring a serious chemical reputation, he was elected a Fellow of the Royal Society in 1769 and published a very successful series of *Chemical Essays* in five volumes, which were reaching their sixth editions by the 1790s.[6] He wangled a better endowment, and attracted large audiences for his practically focused lectures, although chemistry was not part of any formal course at Cambridge. What he was doing would on the Continent have been called 'cameralistics', the science of government and administration; training for those who would become civil servants and who needed to know the economic importance of minerals and other substances. In Cambridge there were many young men destined to inherit estates, who, as the Industrial Revolution developed, wanted to know how to maximise their incomes; Watson's common-sense science fitted the bill.

In his book he expressed the hope that an Academical Institution might be set up to teach young men of rank and fortune the elements of agriculture, the principles of commerce and the knowledge of manufactures. In fact, this was being done in the Dissenting Academies where Priestley and later John Dalton taught, though the good Bishop Watson does not say so; but he sought to do in his lectures what could not otherwise be done in an Anglican institution weighed down by its traditions. Also in the 1780s, Thomas Beddoes in Oxford was attracting large numbers to chemistry lectures – the possibilities for demonstration experiments made chemistry an exciting science to watch, and it could also be smelt and even heard. It is, after all, the science of the 'secondary qualities', colours, tastes, smells and noises – brilliant, pungent, stinking, bubbling, banging and popping. It was said that Beddoes' audiences were larger than those for any Oxford lecturer since the thirteenth century. Beddoes was a medical man, better trained than Watson and more interested in theory, and in the nature of scientific explanation. He was one of the first in Britain to be aware of Immanuel Kant's critical philosophy. He was also politically radical, an associate of Priestley: his support for the French Revolution meant that plans for a government-funded Regius Chair of Chemistry at Oxford for him were shelved; he was fired instead.

The official syllabuses at Oxford and Cambridge had been notoriously old-fashioned, and the oral examinations had become farcical; but with the new century, both modernised their courses, and made assessment more exacting. This meant that 'reading men', serious students wanting a good degree in order to pursue a profession, had no time for frills like scientific lectures – they had to concentrate upon their Classics or mathematics. Only the 'rowing men' (they probably made a row, rather than propelled a boat, but perhaps did both) for whom university was a boozy finishing school, or unambitious and wealthy young men like Charles Darwin who didn't need an honours degree, had the leisure to find their way to scientific lecturers; who therefore, by the 1830s, generally wasted their sweetness on the desert air.

To teach in a university was thus frustrating for the man of science. On the other hand, the rise of industry and commerce was bringing a new confidence that the country's wealth was not limited, and that genuine social progress and economic development was possible. An economy based upon agriculture and handicrafts could only support a few people in leisure or affluence: most people would have to live by the sweat of their brow; digging, delving and spinning like Adam and Eve after their expulsion from Eden.[7] This perception coloured attitudes to education in the eighteenth century, which seem to us very incorrect politically. If most people were going to have to labour all their lives, and if the labour of this great majority could only support a limited number of people in 'white-collar' jobs in the churches, grammar schools and universities, the law and medicine, then it would be both foolish and cruel to educate people above their station, to compete with the sons of professional men, arousing expectations which could not possibly be fulfilled. But if commerce and industry could expand indefinitely, using the new resources of science in navigation and in manufacturing, then an expanding educated middle class would be possible, and indeed essential. Self-help, and institutions for self-education, mutual education and instruction went with the perceived new opportunities of the industrial revolution.

Within Britain, little Scotland was a better-educated country than England or Wales; and many Scots came south to seek their fortune in the eighteenth and nineteenth century, when with an expanding empire the English, Scots and Welsh began to think of themselves as Britons.[8] There was also an influx from poor but well-educated Germany,[9] especially from Hanover, whose Electors from 1715 to 1837 were also Kings of England, Scotland and Ireland. Liverpool in the later eighteenth century was a centre of culture, whose wealthy merchants had the leisure for reading and admiring fine arts; and in the region of Birmingham, opulent manufacturers like Josiah Wedgwood and Matthew Boulton belonged to the Lunar Society.[10] But it was in the grimy boom-town of Manchester that the most famous Literary and Philosophical Society was formed.

The moving spirits were on the one hand clergy, mostly dissenting, and medical men; and on the other nouveau-riche factory owners, who wanted to pick up some culture. Science was a particularly promising part of high culture. To become a connoisseur of painting or sculpture, it was almost essential to have been living among works of art all one's life; anybody whose earlier years had been spent in mean streets, factories and counting houses was unlikely to be able to join comfortably in discussions. Again, the extended vocabulary, easy use of foreign expressions, and awareness of resonances and levels of meaning that goes with literary criticism was not likely to be associated with Mancunian business men in the 1780s. Science was different: astronomy was sublime, botany delightful, mineralogy filled cabinets with precious, beautiful and imperishable specimens. The members of the Lit. and Phil. did not want to exchange information with their commercial rivals; the science they wanted to know about was improving and uplifting, not economically but culturally valuable: they sought intellectual rather than practical understanding, popular science at the appropriate level.

The Society held lectures on a wide variety of topics, and also published a journal in which lectures and research papers were printed. This was rather unusual, though many such bodies did aspire to publication and the 'learned' cachet it brought. In Manchester, John Dalton was employed by the Lit. and Phil. to act as administrator; he also lectured, and did his research on colour-vision, meteorology, gases, analysis and the atomic theory there. The salary was modest; but so were his requirements. He eked it out by taking private pupils, notably James Joule, the wealthy brewer's son, later famous for his work on heat and energy; and late in life he was granted a government 'pension', like those given to needy and meritorious men of letters.

The Manchester Lit. and Phil. was followed by others, notably at Newcastle-upon-Tyne, a booming city based upon the coal industry, where doctors and Unitarians were especially prominent. And then, in 1799, London followed the lead of the provinces: the Royal Institution was founded.[11] Three people brought it into being, and ensured that it would become a great centre of both accessible lecturing and scientific research of the highest quality: Benjamin Thompson (Count Rumford), Banks and Davy. They had a full supporting cast, of aristocracy and landed gentry, and then increasingly as the century wore on, of the upper middle class, professional families and bankers. Whereas in mechanics' institutes and provincial towns mesmerism, phrenology or scriptural geology might be on the menu, the Royal Institution's science was to be of an impeccably establishment character, the scientists' ideal of popularisation, for the well-educated and opulent.

Thompson was an American, a bright and handsome boy who escaped from the drudgery of a shop by marrying a well-off woman some years older than he was. Acquiring land and a commission as an officer, he opted in the American Revolution for the British side. As things turned nasty for such Loyalists, he was hauled before a committee of public safety (an American invention later perfected in France), but escaped to join the redcoats, leaving his wife (whom he was never to see again) and baby daughter behind. Going to England, he managed to get a government post for himself, supposedly looking after the interests of loyalists, and was then commissioned in a regiment he raised (The King's American Dragoons) to fight for the British in the vicinity of New York, where he acquired a bad reputation for ruthlessness among those he perceived as rebels.

When the war ended in defeat, he was not much attracted by the prospect most loyalists faced of resettlement in Nova Scotia or New Brunswick, and so returned to Britain. Crossing the Channel on the same boat as Edward Gibbon, the historian of Rome, he made his way to Strasbourg, where there was a great review of troops. Well mounted, and in his becoming uniform, he caught the attention of Duke Maximilian de Deux-Ponts, whose regiment had fought for the Americans; and who gave him letters of introduction to his uncle, the Elector of Bavaria. The offer of a position at court followed. Hastening back to England, Thompson obtained a knighthood, and permission to enter foreign service, from King George III. Back in Munich, as a colonel, he studied heat and insulation as part of an effort to provide warmer and cheaper uniforms, and then rounded up all the beggars into workhouses to manufacture them. He introduced potatoes into the Bavarian diet, improved cooking stoves, promoted the drinking of coffee, and set garrisons to work growing vegetables for their own use and for the workhouses. He had the famous Englische Garten created in Munich where there had been a royal deer park. The prototype Yankee at a royal court, he was made Minister of War, Minister of Police, Major General, Chamberlain and Councillor of State, and ennobled as Count Rumford (after Rumford, New Hampshire, where he had been married).

He found that when cannon were bored out in the arsenal, the quantity of heat produced by friction seemed endless; and he was convinced that heat was an effect of motion – in opposition to Lavoisier's notion that it was a weightless fluid, 'caloric'. Exhausted by his labours in applied science, and with a great reputation as a philanthropist and rational reformer, he returned to England in 1795. There he spent nearly two happy years advising the nobility about fireplaces and chimneys. But, appalled by the clouds of smoke that hung over London, indicating inefficient use of coal, and the over-large chimneys requiring the use of climbing boys to sweep out the soot, in 1796 he returned in haste to a neutral Munich menaced by the armies of France and Austria. The Elector had fled, but Rumford saved the city by coolly persuading both armies to withdraw and fight it out elsewhere. During the siege, he had invented a portable field-stove, and set up soup-kitchens throughout the city. But his cockiness and prestige made him enemies as well as friends, and in 1798 the Elector sent him back to England as Ambassador.

King George III refused to accept one of his own subjects as a representative of a foreign power, and Rumford looked for a new post – perhaps (despite his record) in the USA at the new military academy at West Point. But in the event, he remained in London where he used his reputation to promote an institution that would advance applied science, bringing to the British state the modernisation he had achieved in Bavaria. In a time of hunger and war, boosting agriculture and manufactures was an attractive prospect; and, in 1799, he succeeded in getting Banks (whose chimneys he had improved) to call a meeting at his house to promote the setting up of a new scientific institution in the metropolis.

Banks had inherited large estates, and at Oxford had become interested in botany; he had then decided (instead of going on the customary Grand Tour to Italy) to sail with Captain Cook to Tahiti, New Zealand and eastern Australia, where it was he who botanised at Botany Bay.[12] After his return, he was in 1778 elected President of the Royal Society, at the age of thirty-four. He held the post until 1820 (his immediate successor, Davy, was born a few days after Banks was elected), thus embarking on a modern-sounding career of scientific administration; building up a learned empire, and supporting useful knowledge, in the service of his country and its more-formal empire – notably Australia. Given the small civil service, Banks' expert advice was crucial to governments.[13] Throughout most of his reign, Britain was at war with France. Science seemed subversive, and it was essential that men of science should hang together. Banks firmly resisted any threats he perceived to the hegemony of the Royal Society, and the unity of science; notably the setting up of specialised societies.

The Royal Society was unspecialised, and by modern standards amateurish. A minority of its Fellows had ever published a scientific paper, and its governing Council never in Banks' time had a majority of active men of science. It was an intellectual club, and the subscriptions paid by the interested gentlemen who joined it paid for the publication of papers by the minority who actually performed scientific research, bringing it its international prestige. The meetings were formal: Banks in court dress was enthroned behind a mace,[14] and papers were read, by their author, or often by the Secretary of the Society. There was no discussion. There were occasionally more informal meetings, 'conversaziones', to which sometimes ladies could be brought; and the hospitable Banks entertained the Fellows regularly. But the Society, and its elegant and austere journal, *Philosophical Transactions*, were not directly concerned with popularising.

Rumford's proposals were for something that would complement the Royal Society (of which he was a Fellow) by doing something different but compatible. As well as a library, it would have a lecture theatre, and a laboratory – the Royal Society had neither. Its object would be scientific philanthropy. Rumford's own interests in stoves and fireplaces extended to machinery of all kinds;[15] he hoped that if artisans and craftsmen could be instructed in science, rather than relying upon traditional methods and rule of thumb ('practice'), then their productivity would increase rapidly. The lecture room, heated by steam, would have a gallery with a separate entrance, like London theatres, where such people could come; while their social betters (whose subscriptions would be the main source of finance) occupied the more genteel seats below. Industrialists would exhibit their latest machines.

This did not work out. Banks and other backers had their eye on the even more urgent business of improving agricultural production, bringing all farms up to the standards of the best and thus forestalling social unrest in years of war and bad weather. As landowners and churchmen, they were not especially enthusiastic about dissenting manufacturers; who equally did not wish to display their latest improvements for everyone to see – industrial espionage, by domestic and foreign agents, was a genuine problem, and patent legislation expensive and uncertain. Rumford, with his amazing capacity to ingratiate himself with the mighty and his self-confidence, did not delight everybody, and his administration seemed both autocratic and informal. His first choice for the important professorship of chemistry, that ubiquitous, necessary and revolutionary science, was Thomas Garnett.[16] He had become eminent for his analyses of mineral waters, an important activity when Bath, Tunbridge Wells, Harrogate and other spa towns were at the height of their popularity and medical reputation; but he was unhappy in London, and his lectures failed to hold an audience requiring to be entertained. Thomas Young also lectured, on physics,[17] but these were too much like a high-powered academic course to go down well. In the autumn of 1800, Rumford began negotiations with a young man in Clifton; and, in January 1801, at the age of twenty-two, Davy joined the staff of the Royal Institution. His lectures and research would set its course, and put some London science into the same league as Parisian.

With the peace in 1802, Rumford went to Paris, and remained there even after the war broke out again. He married Mme Lavoisier and lived unhappily ever afterwards. But Davy's inaugural lecture in that year was an enormous success. Because access to the Continent was closed, those in search of culture had to find it at home; and, with a sophisticated audience very different from those in Manchester or Newcastle, Davy found that he could make science seem an important component of high culture. He emphasised how economic and social progress depends upon inequality, which must have reassured his affluent audience; and he got across the excitement of scientific investigation. Sometimes, he even did his research in public, with a bank of seats in the laboratory where onlookers could perceive his creative misuse of apparatus in bursts of inspired activity. Later, he sadly reflected that he had failed to make the nobility and gentry actually take up science to any extent, but that is what one might have expected. Noblemen and their ladies did not, as a rule, actually play musical instruments, paint pictures or write novels seriously – their role was as patrons of those who did. In the sciences, the immensely wealthy and reclusive Hon. Henry Cavendish had been the great exception;[18] and he was not interested in popularising.

Before lectures, Davy dined lightly upon fish, and he went through the experiments carefully with his assistants. Such demonstrations, looking spontaneous and drawing the audience in as witnesses, became a feature of the Royal Institution, and indeed of scientific lecturing in Britain – where hands-on science teaching came later than in Germany. Some of Davy's were dramatic, as when he had a hollow clay cone filled with potassium and poured water into it. Volcanoes are mostly near water and Davy's conjecture was that there might be beneath them deposits of active metals, which would produce an eruption as his model did. Later he demonstrated a safety lamp by plunging it into a vessel of explosive gas. There are manuscripts of his lectures, and he published a syllabus for each course; but he had the advantage that he could talk about those aspects of the science that interested him. This is very different from taking classes of students through all aspects of a discipline, in order that they may pass examinations and enter a profession; and it is one of the attractions of the popular lecture. All lectures should arouse enthusiasm, but it is more difficult if the lecturer does not feel it.

Hearers reported Davy's bright eyes; Laetitia Barbauld exhorted readers[19] of her poem '1811' to:

> Call up sages whose capacious mind
> Left in its course a track of light behind;
> Point where mute crowds on Davy's lips reposed,
> And Nature's coyest secrets were disclosed.

Clearly these were not lectures read out from a manuscript. Davy, like most of the great nineteenth-century science lecturers, must have spoken extempore, or possibly with a note of a few headings. There is no doubt that performances of this kind are much more effective as oral communication than most academic talks or other speeches given today. We no longer trust our memories, and lose the consequent rapport with listeners (and perhaps the affinities with tight-rope walking) that was crucial in the effectiveness of Davy, Faraday, Huxley and Tyndall[20] (who gave 430 lectures to varied audiences, in 'a rare lucidity of style and a rich drapery of language'). Professing should be a performance art. Science and theatre are not wholly separate.[21]

Davy did write a textbook, or rather the first volume of one, but it was not a success. The art of the textbook is very different from that of the popular lecture series. The whole discipline must be covered in a logical and systematic way, and the special research interests of the author must not be allowed to dominate the discussion. Davy's text was mostly concerned with his own discoveries, which had already, and more appropriately, been published by the Royal Society as scientific papers – in a much more expansive and accessible form than happens nowadays, but nevertheless addressed to a scientific rather than general reader. But for lectures, current research can be an excellent topic. We know that Davy danced about the room in ecstatic delight when he isolated potassium on 19 October 1807; his delight must have come through to the audiences for whom he subsequently repeated the experiment. Fragments of metal caught fire and flew through the air, making bright coruscations; the front row were very near the lecturer's table, safety precautions in the Royal Institution were few, and the effect must have been stunning. Although large, the Royal Institution's semi-circular lecture theatre seems intimate; it is steeply raked, the audience is on three sides, the big table occupies the centre and there is a U-shaped recess in it so that the lecturer actually stands amid his equipment at the focus of all eyes. Davy competed successfully with all the other theatrical activities in London, attracting and holding an audience. His bright eyes caught those of his audience, who sent him not only commissions to analyse rocks or soil but also fan mail and invitations to dinner parties and salons.

Davy's early work at the Royal Institution had been on tanning and agriculture, where he sought to show how traditional activities could be understood and improved by the application of science. Essentially his lectures and papers vindicated the best practices at which common sense had arrived. In a retort with its spout inserted under a little patch of turf, he put some manure. As the manure rotted, over the course of a week or two, so the area round the spout became much greener than the rest. This smelly, common-sense experiment established for all who saw it that leaving manure to rot is a bad practice because the nutritious ammonia escapes into the air rather than getting into the soil. Finding substitutes for oak bark, by then already in short supply when leather was the predominant plastic material, Davy came to lecture one day with one shoe tanned with oak, and the other with the Indian shrub, catechu. Audiences loved this accessible, practical science and, as well as being invited to the Duke of Bedford's sheep-shearing at Woburn (where his health was drunk, and where he supervised experiments on fodder grasses), Davy was in demand socially, as a handsome, eligible and intellectual bachelor whose conversation, like his eyes, would sparkle. He made science fashionable.

The work on tanning, applied chemistry of economic importance, had brought him the Royal Society's Copley Medal – its highest award. But in the autumn of 1806, after his holiday and before the London season began in November, Davy returned to earlier galvanic studies,[22] which would earn him a medal from the French. He had consistently believed that the electric battery that Alessandro Volta had announced in 1799 – an alarm bell for experimenters – depended for its action upon a chemical reaction rather than mere contact as Volta and others supposed. When the wires from the battery, or 'pile', were put into water, bubbles of oxygen and hydrogen arose from it; Davy believed that electricity was what bound elements together in compounds, and could be used in analysis. This theoretical insight (worked out in experiments involving apparatus of silver, gold, platinum and agate) won him the French prize. He expounded it in lectures, and the following year used it in analysing potash.

Jane Marcet, a doctor's wife, attended Davy's lectures, and in 1807 published *Conversations on Chemistry*, directed especially at girls who had, like her, been dazzled:[23]

> On attending for the first time experimental lectures, the author found it almost impossible to derive any clear or satisfactory information from the rapid demonstrations which are usually, and perhaps necessarily, crowded into popular courses of this kind. But frequent opportunities having afterwards occurred of conversing with a friend on the subject of chemistry, and of repeating a variety of experiments, she became better acquainted with the principles of that science.... She perceived, attending [Davy's] lectures the great advantage which her previous knowledge of the subject ... gave her over others.... Every fact or experiment attracted her attention, and served to explain some theory to which she was not a total stranger.

Faraday, the bookbinder's apprentice given a ticket to a lecture-series a few years later by a well-wisher, prepared himself by reading Jane Marcet's book (and then wrote up his notes, presenting them to Davy with a request for a job). Clearly she was right: lectures on their own cannot give the solid basis required for someone who really wants to master a field. Demonstrating where the grass has become greener is one thing, but getting across the subtle relationship of fact and theory that constitutes genuine science is another. An informed audience could not really be expected at somewhere like the Royal Institution; but Davy had the gift of pleasing everyone.

After isolating the extraordinary metal potassium, which floats on water and bursts into flames as it decomposes, Davy could glow with patriotic fervour. London was going beyond Paris in a major part of chemistry. Lavoisier had thought that potash might be a compound, but he did not know what its constituents might be. His 'oxygen' was so named because he believed that it was a component of all acids. After investigating potassium, and proving that the strong alkali caustic potash contained a lot of oxygen, Davy went to work on acids; and, in 1810, established to his own satisfaction (and ultimately to everybody's) that not only the weakly acidic bad-eggs gas hydrogen sulphide, but also the very strong acid from sea salt, contained no oxygen. The green and choking gas associated with the latter should not be called 'oxymuriatic acid', as Lavoisier's associate C.L. Berthollet had done, but given a new, theory free name from its colour, 'chlorine'. Davy exulted:[24]

> The opinions of Berthollet have been received for nearly thirty years; and no part of modern chemistry has been considered so firmly established, or so happily elucidated; but we shall find that it is entirely false – the baseless

fabric of a vision.... The confidence of the French enquirers closed for nearly a third of a century this noble path of investigation, which I am convinced will lead to many results of much more importance than those which I have endeavoured to exhibit to you.

'Baseless fabric' is from Prospero's great speech in *The Tempest*; Davy's audiences could be expected to respond appropriately to tags from Shakespeare.[25]

This was from one of Davy's last series of public lectures, for in 1812 he was knighted, and married a wealthy bluestocking widow, taking early retirement. He wrote that he would have more time for research, and indeed in 1815 did invent the miners' safety lamp; but travelling widely, and then from 1820 as President of the Royal Society, he found himself, like other middle-aged eminent scientists, moving willy-nilly into administration. His career seems in fact to show that having to explain research in public is very stimulating and that research and teaching really cannot or should not be separated. He had helped to make chemistry exciting and popular. In a poem of 1820 by Henry Luttrell (concerned with London fog) we find:[26]

> O Chemistry, *attractive* maid,
> Descend in pity to our aid!
> Come with thy all-pervading gasses,
> Thy crucibles, retorts and glasses,
> Thy fearful energies and wonders,
> Thy dazzling lights and mimic thunders!
> Let Carbon in thy train be seen,
> Dark Azote and fair Oxygene,
> And Woolaston, and Davy guide
> The car that bears thee, at thy side.

Contemporaries contrasted William Hyde Wollaston's determination to avoid error with Davy's more romantic urge to find truth – which is more fun in a lecturer. At the Royal Institution, Davy's mantle passed to the much more prosaic W.T. Brande, with the active assistance of Davy's protégé, Michael Faraday, whose enthusiasm for chemistry had been kindled by his getting a ticket to hear those lectures of Davy's on chlorine. Brande's real forte was textbooks, explicitly accompanying his courses aimed at London's medical students (for whom, after the Apothecaries Act of 1815, formal instruction as well as apprenticeship was required).

In 1820, Davy became President of the Royal Society, and had to make a speech each year at the AGM on St Andrew's Day, 30 November. This involved a review of the year, and often some brief obituaries: the French Academy of Sciences had long made a habit of publishing elegant eulogies of members.[27] At the Royal Society there were also medals to distribute: the Copley medal was an old one; Rumford had endowed another, named after him; and, through Robert Peel, Davy secured two annual Royal medals – so there were several people, dead and alive, to notice. Davy was proud enough of his rhetorical achievements in thus boosting British science to have them collected together and printed in 1826 at the Society's expense.[28] They clearly were a kind of popular lecture, though addressed to one of the world's leading scientific societies. They functioned to help the amateurs understand the significance of what their more active brethren were up to; and also to inform emerging specialists, making mathematicians appreciate some botany or geology, or vice versa. Charles Babbage, irascible as always, denounced the publication as corrupt.[29] Davy was taken ill at just that time, the book did not sell well, but the discourses are good examples of scientific rhetoric, getting the establishment's version of science across to the reading public. Davy had risen to the Presidency from humble origins, like his contemporary Sir Thomas Lawrence, President of the Royal Academy, who painted the splendid swagger portrait of him now at the Royal Society; he had to work on getting ceremony right, unlike Banks, a gentleman born, who did not have to worry about the impression he was making.

Meanwhile, at the Royal Institution, Faraday was taking over with, and then from, Brande the position Davy had occupied. Liberated from relatively routine science after Davy's death in 1829, he went on to consolidate it with even more important researches into electrochemistry, and then electromagnetism. Marrying for love rather than money, he spent his whole career lecturing at the Royal Institution, where his many researches in chemistry and electricity made for wonderful lectures – including one in which a poker and tongs were flung up in the air to be captured by a large electromagnet above the heads of the audience, and were followed by the coal scuttle. He inaugurated the Christmas Lectures for children, and his series on *The Chemical History of a Candle* is an enduring classic of popular science by one of the greatest of all scientists. They were taken down in shorthand, and thus retain some of the freshness of the spoken word: 'read until you hear the voices' is a maxim very applicable to published lectures.

Prince Albert brought the royal children along to hear Faraday, and ever since Faraday's time these lectures that he began have been a source of delight to young hearers, who are introduced to current science by someone who is both an expert and a good communicator. Such people are rare, but in the nineteenth century, without radio and television, people had to be content with what they could get, turning out for sermons and lectures and finding inspiration in what must often have been unreliable or just plain dull. Queen Victoria had evidently attended the lectures on the four ancient elements, fire, air, earth and water, delivered by Thomas Griffiths, a chemist attached to St Bartholomew's hospital in London,[30] which when published he dedicated to her by permission, and with 'fervent loyalty'. He dilated upon the goodness and wisdom of the Creator; stressed how experimental a science chemistry was, and illustrated the talks with frequent demonstrations. Similar lectures must have been given by medical men to audiences up and down the country, rather less august, but equally keen to be informed and entertained; and, in the USA, to sophisticated audiences in Boston, New York and Philadelphia, and to those avid for self-improvement further west. John Scoffern, who edited some of Faraday's lectures for publication, published a little book purporting to be a series of lectures delivered by an Old Philosopher to a youthful audience in Devonshire.[31] Fully illustrated with woodcuts, so that the experiments could be repeated, and recording the coughing which greeted the generation of chlorine, the little book shows how lectures, real or imaginary, could be much more lively than a textbook, and yet get quite a lot of science across – the published version solidifying the experience or illusion of hearing.

In provincial cities, resident or itinerant lecturers could find audiences keen to learn more about exciting, and maybe useful or edifying, discoveries in the sciences. John Phillips, based in York, also lectured on geology in various venues in Manchester, Liverpool, Chester, Newcastle and Bristol.[32] His lectures were authoritative, up-to-date and mainstream; but there were also very popular lectures on phrenology, or cranioscopy, the science of demonstrating character through the shape of the head. This would-be science, originating in Vienna, was brought to Britain by J.G. Spurzheim, who published his lectures in a large volume, illustrated with diagrammatic heads and with portraits.[33] Phrenology became a craze; to its devotees, it seemed an important part of medicine, and in Edinburgh a phrenological society was formed and published a journal. The would-be science appealed to those seeking release from establishment views in religion and science: the determinism, attributing character to brain development and thus to skull shape, was attractive in the same way that attributing it to genes is today. We may perhaps see phrenology as the psychoanalysis of the nineteenth century. It was never fully respectable, but it entered into the thought and language of the era, was a serious matter for educationalists and made a splendid topic for lecture-demonstrations.

The Royal Institution, where phrenology was not welcome, had been designed to accommodate 'working men' in its gallery, with its separate entrance as in other theatres; but, in the event, little seems to have come of this attempt to integrate social classes through their interest in science. But artisans, the skilled workers rather than the labourers of the day, were keen to learn, and gas-light made evening classes practicable. In Mechanics' Institutes, as in the more up-market and middle-class Lit. and Phils, programmes of lectures were organised, either by the men themselves or by their paternalistic employers; and such places also assembled good libraries as, with new technology, the price of books fell sharply in the second quarter of the nineteenth century.[34] Sometimes 'working men' could benefit by getting their popular science from active and eminent men of science. Just as prominent doctors in cities and in the provinces acquired prestige and standing by charitably giving some of their time at a clinic for the poor, so prominent natural philosophers would give a lecture or a course for working men. This might be associated with a meeting, for example of the British Association for the Advancement of Science, or with an institution, such as the Museum of Practical Geology in Jermyn Street, London. Here, the anatomist Richard Owen and the physiologist and palaeontologist Thomas Huxley held forth, somewhat competitively. Museums indeed were important loci for lectures on science as well as humanities in the nineteenth century, with exhibits on hand to help make points.[35]

Huxley enjoyed lecturing to working men more, so he said,[36] than holding forth at the Royal Institution, or in the formal courses he gave in South Kensington in what was to become, after his death, a part of Imperial College. Apart from the very occasional Faraday or Alfred Russel Wallace, artisans could not realistically aspire to become men of science, supporting themselves by their teaching and research. For them, science would be a part of high culture. Huxley's lecture of 1868 on a piece of chalk is a tour de force in this genre.[37] His technique was to demonstrate that science was trained and organised common sense (the adjectives are important), by taking the audience through evidence and encouraging them to reason about it – rather like a skilled lawyer guiding a jury towards their verdict. The chalk is composed of fossil remains, and Huxley used this to evoke vividly past epochs of the Earth's history: but his main topic was crocodile fossils. Over a very long period of geological time, different species of crocodile have succeeded one another; they are different, but not very different. Why should this be so? Huxley devoted the last 10 per cent of the lecture to interpretation. He invited his audience to make up their own minds:

> Choose your hypothesis; I have chosen mine. I can find no warranty for believing in the distinct creation of a score of successive species of crocodiles in the course of countless ages of time. Science gives no countenance to such a wild fancy; nor can the perverse ingenuity of a commentator pretend to discover this sense, in the simple words with which the writer of Genesis records the proceedings of the fifth and sixth days of the Creation.

For Huxley, the operation of natural causes underlies all that we see – and to be an agnostic[38] (a word he coined), suspending judgement as long as possible, was an important aspect of scientific method. But here he believed that the time had come for his audience to reach a verdict.

His peroration was splendid, invoking the limelight of the theatre:

> A small beginning has led us to a great ending. If I were to put the bit of chalk with which we started into the hot but obscure flame of burning hydrogen, it would presently shine like the sun. It seems to me that this physical metamorphosis is no false image of what has been the result of our subjecting it to a jet of fervent, though nowise brilliant, thought tonight. It has become luminous, and its clear rays, penetrating the abyss of the remote past, have brought within our ken some stages of the evolution of the earth.

Huxley was a prominent member of the London School Board, set up when elementary education became compulsory in Britain in 1870; and he pressed for science, and particularly scientific method, to be included in the syllabus, along with the 'three Rs' – reading, (w)riting and (a)rithmetic. He was also both a member and a key witness at the Royal Commission on Scientific Instruction and the Advancement of Science, chaired by the Duke of Devonshire, which reported in 1872[39] in two large and stout volumes – a wonderful guide to all that was going on. Huxley was resolute in his belief that science should be a prominent part of education, but that 'two cultures' must be avoided.

Huxley thus turned around the idea that religion was the clue to making science palatable, by using irreligion, which worked just as well, and in the same way.

It might be noted that agnosticism was made respectable in part because of a very different set of lectures, the 'Bampton Lectures' delivered regularly in the University of Oxford, and in 1858 by the logician Henry Mansel.[40] There had always been an austere *via negativa* in Christianity, declaring what God was not like rather than the other way round; and Mansel, making use of Kantian philosophy, was in that tradition. His book, which was well received, led to his appointment to professorships at Oxford, and in 1868 to the Deanery of St Paul's Cathedral in London. This was unlike the notorious volume of essays by Oxford contemporaries including Benjamin Jowett, *Essays and Reviews*, which introduced the public to Biblical criticism, and the difficulties of reconciling miracles and the *Genesis* story with the science of 1860[41] – the *Origin of Species* had just appeared. But that is another story, and meanwhile we continue our focus on communication by looking at the written language of science as it developed over the long nineteenth century.

4 Poetry, metaphor and algebra

Antoine Lavoisier's *Elements of Chemistry* is approachable, and thus quite different from Isaac Newton's *Principia Mathematica*, or Carl Linnaeus' *Systema Naturae*, but more like Charles Darwin's *Origin of Species*. Some of the great original works of science have been intended for experts only (Copernicus even had a deterrent Greek slogan, 'non-mathematicians keep out', on his frontispiece); but others were meant for a much wider readership, indeed for public as well as professional understanding and enlightenment. In Linnaeus' book, the focus is on both observation and language: exact and minimal descriptions, in a botanical Latin that is still going strong after two and a half centuries; and two-word (binomial) scientific names that are internationally accepted. The language[1] abounds in adjectives, and lacks the sonorous and elegant features of Cicero's or Virgil's Latin; but the names can call up associations, as *Tyrannosaurus rex* does among the dinosaurs' names, for those who know a little Greek and Latin. The plant names, following Linnaeus' use of sexual characters of flowers in his classification, struck the prudish Victorian critic John Ruskin as lewd;[2] but this could make systematic botany more fun for less-inhibited mortals.

That thought struck Erasmus Darwin, a physician in the little cathedral city of Lichfield and subsequently in Derby. He was a polymath, interested in everything,[3] and a prominent member of the informal dining club calling themselves the Lunar Society of Birmingham[4] because they met at the full moon, and could thus find their way home. Other members of this distinguished group included Josiah Wedgwood, the potter, James Watt of steam-engine fame and Joseph Priestley. Darwin turned to verse to get the new botany across to the public: identifying plants was no longer a matter of connoisseurship, but was systematic, could be readily learned, and might be entertaining. Instead of the multiple and uncertain criteria of older systems, Linnaeus' names for flowers were based on counting the sexual parts, so that *Triandria Digynia* would be the group which contained three stamens and two pistils. Within the group, plants would have a double-barrelled name, the first word denoting the genus, and the second the species. Hoping, as he put it,[5] to 'inlist Imagination under the banner of science', Darwin's couplets abound in classical allusions to nymphs and dryads, playing also with the Latin names. He wrote, for example, about poisonous plants (with an allusion to *Hamlet*):

If rests the traveller his weary head,
Grim MANCINELLA haunts the mossy bed,
Brews her black hebenon, and stealing near
Pours the curst venom in his tortured ear. –
Wide o'er the mad'ning throng URTICA flings
Her barbed shafts, and darts her poison'd stings.
And fell LOBELIA with contagious breath
Infects the light, and wings the gale with death.

Abundant footnotes, 'interludes' where the poet and the bookseller exchange views about the odd exercise of popularising science in comic verses, and additional endnotes (particularly about the dreadful upas tree, poisoning everything for miles around) make the text unlike other scientific books. Moreover, the second volume was published before the first, on the Economy of Vegetation, based upon the ancient elements of fire, earth, water and air. That is an overview of the whole realm of science, from the solar system through steam engines to geology (and the evils of slavery, via mining), the circulation of water, and the winds, before ending up with a renewed focus upon botany ready for the second volume and its Linnean theme. In the eighteenth century, the 'Enlightenment', Newton's mathematical physics was augmented by the equally important emphasis upon classification and exact language in natural history. Now that could be made popular and fun to learn.

Erasmus Darwin's language was exuberant rather than exact, but he knew his science. In his later *Temple of Nature*[6] he sketched an evolutionary history of the world, revealing himself as a polymath, at ease and authoritative across the range of sciences from astronomy through to natural history and physiology, and from the beginning of things to the present. The poem begins:

By firm immutable immortal laws
Impress'd on Nature by the GREAT FIRST CAUSE.
Say, MUSE! How rose from elemental strife
Organic forms, and kindled into life;
How Love and Sympathy with potent charm
Warm the cold heart, the lifted hand disarm;
Allure with pleasures, and alarm with pains,
And bind Society in golden chains.

Love and strife had been the great antagonist powers in the cosmology of the ancient Greek philosopher Empedocles,[7] where their clash gave rise to new syntheses; and their dialectical opposition gave Darwin the opportunity to account for the living world, mankind and society, leading up to Richard Arkwright's factories which were bringing the modern industrial world into being. His world was begun by a distant and benevolent Deity, whose existence Darwin allowed himself elsewhere to doubt,[8] while 'Nymphs enraptured', goddesses and seraphs in which he certainly did not believe kept the action moving. Darwin's progressive, optimistic and evolutionary picture, where due place was nevertheless given to violence, rapine and disease, pulled together a great deal of the scientific and technical knowledge of the day:

> The wolf, escorted by his milk-drawn dam,
> Unknown to mercy, tears the guiltless lamb;
> The towering eagle, darting from above,
> Unfeeling rends the inoffensive dove;
> The lamb and dove on living nature feed,
> Crop the young herb, or crush the embryon seed.
> Nor spares the loud owl in her dusky flight,
> Smit with sweet notes, the minstrel of the night;
> Nor spares, enamour'd of his radiant form,
> The hungry nightingale the glowing worm;
> Who with bright lamp alarms the midnight hour,
> Climbs the green stem, and slays the sleeping flower.

Once again, the footnotes enlarge upon the poem, explaining in prose just what is alluded to, and bringing the reader up-to-date with galvanism and other new discoveries. Darwin's popularising was very successful, making him probably the best-known poet of the 1790s: it was just right for its time, odd though that may seem from our perspective. Darwin was famous enough to be a major target for parody in the right-wing *Anti-Jacobin*, in the 'Loves of the triangles' written by William Pitt, Hookham Frere and George Canning.[9] Darwin's success depended upon the use of metaphor backwards and forwards between scientific and ordinary language: Linnaean language was richer than it had seemed.

Priestley's science of chemistry had evolved from alchemy,[10] where metaphor expressed pictorially[11] and in texts was crucial – though the poems might often be written to tantalise all but adepts, blinding them with science rather than instructing or enthusing them.[12] Thus pictures and texts had layers of meaning, and the parallels between self-improvement and transmuting base metal into gold were clear. George Herbert used alchemical imagery in his famous poem, 'The Elixer', now often used as a hymn, written in 1653, where one stanza, comparing God's blessing to the effects of the philosopher's stone, and referring to the 'tincture (for thy sake)' goes:[13]

> This is the famous stone
> That turneth all to gold
> For that which God doth touch and own
> Cannot for less be told.

Alchemical themes can be detected behind a number of literary works of that period, more or less explicitly; and scholars have disagreed about whether some texts are recipes for chemical processes, or are metaphorical. The psychiatrist Carl Gustav Jung saw alchemical archetypes occupying an important place in our unconscious minds,[14] and coming up in his patients' dreams. Though not many have followed him that way, its possibility indicates how open to lateral thinking alchemical language was. When chemistry, in the form of pharmacy, was applied to medicine, it brought a curious baggage of nomenclature and of symbolism, running far back into the ancient world and including astrological signs as well as Arabic terms, and people's names. This, though picturesque, did not make the science easy to remember or understand. It also had an aspect of bondage to authority and tradition, from which the natural philosophers of the seventeenth century had struggled to free themselves.[15]

In the mid-eighteenth century, Pierre Joseph Macquer in his *Elements of Chymistry* remarked that:[16]

> Chymistry became an occult and mysterious science; its expressions were all tropes and figures, its phrases metaphorical, and its axioms so many enigmas; in short, an obscure unintelligible jargon is the justest character of the Alchymistic language.

Macquer, an Academician since 1745, wanted chemistry to become a simple science founded upon facts: as well as his textbook, he wrote a *Dictionary* of the science,[17] the first in what became an important genre, though it seems nowadays a curious way of getting science across. One of his pupils was the Marquis de Condorcet, 'philosophe', logician and man of science, and an important political figure in the early years of the Revolution. Macquer saw his science emerging from profound obscurity, and gaining the admiration of the world as it appeared in open day. Ingenious men, he declared, vied with each other and communicated their discoveries, so that chemistry and the 'arts' that depended upon it made rapid progress:

> In a word, it put on a new face, and became truly worthy of the title of Science; founding its principles and its processes on solid experiments, and on just consequences deduced from them.

Macquer went on to applaud the way that chemists were daily increasing useful knowledge; and he compared the science to geometry (long the great example of deductive logic) for its ample field of inquiry, and its wide usefulness.

In response to Macquer and in imitation of Linnaeus, the French chemist Guyton de Morveau had begun to devise a more exact and clear language for chemistry. His enterprise got partly taken over by Lavoisier, well-placed, wealthy and ambitious; and, with Claude-Louis Berthollet and Antoine François Fourcroy, they proposed their new system to the Academy of Sciences on 18 April 1787.[18] This formed the basis of the language of chemistry still in use, with terms like 'lead sulphate' and 'ammonium nitrate'; and, though it may nowadays seem rebarbative to children who have to learn it, it was an important part of the appeal of chemistry to wide audiences in the years around 1800.

For Lavoisier, it combined with his new theory of burning, in which something (oxygen) was absorbed from the air rather than emitted into it (phlogiston, from the Greek meaning flammable), and made his transmutation of chemistry possible. His book appeared in 1789, the year the Bastille fell, when it was bliss to be alive, and to be young was heavenly as the political and social world was turned topsy-turvy. He was the first scientist to claim that he too was making a revolution, an intellectual one.[19] In the event, the political revolution he had supported brought him little joy – he was guillotined in 1794. Lavoisier's work improved understanding of chemical reactions, even though practice was not much changed: it consolidated the status of chemists as men of science, now distinct from pharmacists who were tradesmen and whose science was 'professional' and thus not suitable for ladies and gentlemen: chemistry was becoming 'polite', in France, Scotland and England.[20]

Lavoisier meant his book[21] to be widely read, improving public understanding of science generally, as well as chemistry in particular: it might be called semi-popular. Macquer had invoked geometry, and in the hands of John Dalton and then of organic chemists, the science was indeed to become applied geometry, so that the DNA spiral is one of the icons of our day. But in his preface, Lavoisier appealed to algebra as an exact language, free from metaphor or imagery: where geometry is suggestive, algebra is austere (and to many, daunting). Chemistry being an empirical, experimental science, his book could not be like Euclid's, a chain of reasoning where each link supported those beyond it: he had to go inductively, from carefully established facts. The student must not be distracted into system-building by a rich and flexible vocabulary. Some have always seen imaginative leaps, 'Eureka moments', as an essential part of science, but for Lavoisier it was the realm of cool reason. 'Flowers of sulphur', 'oil of vitriol', 'sugar of lead' are evocative, as are 'realgar', 'orpiment' and 'Glauber's salt': but these names gave, at best, few or misleading clues about composition, and might set the tyro off on altogether the wrong track. Linnaeus' language had shown itself in Darwin's hands to be sexy and funny, while algebra, and Lavoisier's chemical language, were serious and free from nuances:

> From the name alone may be instantly found what the combustible substance is which enters into any combination; whether that combustible substance be combined with the acidifying principle [oxygen], and in what proportion; what is the state of the acid; with what base it is united; whether the saturation be exact, or whether the acid or the base be in excess.

But no more than these things would be suggested. In the belief that the new language would greatly assist the beginner, Lavoisier quoted a letter from Tobern Bergman, the famous Swedish chemist who had recently died, to Guyton: 'Spare no improper names; those who are learned will always be learned, and those who are ignorant will thus learn sooner.' But it was not obvious that such algebra would make chemistry popular among those who had no professional reason to learn it; though, in the event its novelty and the evidence that the science was both undergoing profound and rapid change, and had a new language putting everyone on a level, made it accessible and fun to watch and practise.

To some, chemistry could still seem poetic, its logic stimulating Friedrich Schlegel around 1800 in writing 'fragments' which cohere into a whole;[22] while Coleridge went to Davy's lectures to improve his stock of metaphors. These Romantic appropriations of the science were not what Lavoisier had in mind; and, with Erasmus Darwin, the epoch of popularising science in verse came to an end. Science would be expounded in prose, but that did not mean that it had inevitably to be prosaic. Lavoisier hoped his readers would repeat his experiments, and be convinced thereby of his theory. Britons were usually more cautious. Watson in his evidently-lively lectures at Cambridge had emphasised usefulness rather than theory, and so did his successors, Francis Wollaston and Smithson Tennant;[23] Priestley in his writings had been scornful of hypotheses, and enthusiastic in presenting new facts.[24] William Nicholson translated texts from the French, and also took up Macquer's idea of compiling a dictionary of chemistry.[25] There, and in his textbook, he adhered to a Baconian empiricism, seeing the new language as optional, and preoccupation with naming and theorising as a diversion from real science:[26]

> With regard to nomenclature and theory, I have attempted to keep clear of every system. I have called things by such names as are most in use, except where the usual name pointed too evidently at theories either long-since exploded, or not yet proved: and in the relation of facts I have found it much less difficult to exclude theoretical allusions than I at first apprehended, when I formed the determination of confining the theory, for the most part, to the ends of chapters.... As I think the antiphlogistic hypothesis equally probable with the modified [phlogiston] system of Stahl ... I have judged it proper to explain both. And this I have endeavoured to do in such a way, as to create in the chemical student an habit of steadily and calmly attending to the operations of nature; instead of indulging that hasty disposition for theorizing, which indeed might pass, on account of its evident impropriety, without any earnest censure, if we had not the mortification to see it too much practised by men entitled to the best thanks of the scientific world, and on that account possessing greater power to mislead.

He evidently saw no difficulty in translating between the old and the new; and pharmacists seem to have been with him, because 'anti-phlogistic' remedies to reduce fever went on being prescribed for a generation (to the ailing Davy in the late 1820s, for example), and survived in veterinary prescriptions well into the Victorian era.

Thomas Kuhn's powerful vision of normal science as dogmatic, based on a 'paradigm' or example taken from some great practitioner, electrified my generation when he presented the idea at a conference in Oxford in July 1961:[27] his subsequent book gave the public a new way of understanding science. He recalled to graduate students like me who had recently finished a science degree, and then moved across into history or philosophy of science, the world they had lost. There, facts and interpretations had to be learned, and experiments got right: coming up with a different figure for the melting point of a phenylhydrazine derivative from the one in the book was not an interesting discovery, but simply a sign of incompetence deserving a low mark. The same would go for deducing a different molecular structure for it from the one accepted. There was a royal road to truth, and no time for trial and error in the hope that students might stumble upon it: their feet must be set on the way that they should go. Friends of mine, graduate students in science, were indeed, in doing their research for the PhD, filling in a picture of which the outlines seemed pretty clear: new worlds were not opening up to their astonished gaze.

For Kuhn, anomalies would be sure to build up as the picture failed to match reality; but, as long as possible, the appearances would be saved by ad hoc hypotheses – as when phlogiston seemed to have negative weight. Eventually, a Lavoisier would appear and set off a revolution by taking a different perspective: if he could convince the rest of the scientific community (and Lavoisier's wealth and social position, his role in the academy of sciences and his control of the journal *Annales de Chimie* ensured this), then his new paradigm would prevail. Revolutionaries struggled with questions difficult to ask; their successors with questions difficult to answer. A new language, or at least old words taking new meanings, was an important part of such a Kuhnian revolution.

Translation is always problematic, even hilarious, as we all know from guide books and instruction manuals: it should be reasonably straightforward for technical or scientific prose, but for poetry (especially from a very different epoch and language), it cannot be exact or definitive – it must involve some betrayal of the author's nuances, though there are said to be translations better than their originals. For Kuhn, paradigms involved different world-views, and adequate translation between them was impossible. Contemporaries of Lavoisier like Nicholson did not see it that way. To them, it looked more like having to convert between the new-fangled French kilograms and litres, and the scruples, grains, ounces, pounds, cubic inches and gallons to which they and their readers were accustomed.

Priestley could not bring himself to accept 'oxygen' and the rest of the system, which he saw as complex and theory-laden, akin to Descartes' idea that planets were carried round the Sun in whirlpools or vortices of subtle matter – a notion happily exploded by Newton.[28] He was also hostile to jargon, and would have been bitterly disappointed that his favourite chemistry was, by the twentieth century, being taught as dogmatic 'normal science': his hope was that it would be accessible to all, as it had been to him, that everyone interested would take up experimenting, and that scientific truth, defeating kingcraft and priestcraft, would make the people free. It is a curious twist of history that has left the chemical writings of Priestley (who delighted in clarity) hard for us to follow because they abound in terms like 'dephlogisticated air', whereas those of Lavoisier (who sought a kind of algebra) are much more accessible because his language became the modern one. But Priestley was serious-minded, not tempted to write verses.

Priestley was however delighted with the young Davy's book on laughing gas (and the other oxides of nitrogen) even though he used the French terminology. He wrote to him from his American exile on 31 October 1801:[29]

> It gives me peculiar satisfaction, that as I am now far advanced in life, and cannot expect to do much more, I shall leave so able a fellow-labourer of my own country in the great field of experimental philosophy. As old an experimenter as I am, I was near forty before I made any experiments on the subject of *air*, and then without, in a manner, any previous knowledge of chemistry.... I rejoice that you are so young a man, and perceiving the ardour with which you begin your Career, I have no doubt of your success.

Davy was in Bristol, working with Thomas Beddoes at the Pneumatic Institution, funded by Wedgwood and with apparatus designed by Watt, investigating the medical value of the gases Priestley had isolated. He would indeed have a successful career, becoming a baronet, an apostle of applied science, and President of the Royal Society: but by virtue of accepting patronage and then responsibility, and compromising the left-wing purity or intransigence that Priestley, Beddoes and Darwin had displayed. In Bristol, there were also the poets Coleridge and Southey, who had hoped to join Priestley on the banks of the Susquehannah, bringing with them a utopian community, or 'pantisocracy'. Wordsworth had also been there for a time (Beddoes had sent him and Coleridge to visit Germany as the new intellectual centre of things), and the memoirs of Joseph Cottle, the publisher of *Lyrical Ballads*, 1798, give a splendid picture of their milieu.[30]

Unlike Priestley and Lavoisier, Davy strongly felt the urge to write poetry, and we have a good deal of it, though he never published a volume of verse; all his biographies contain some, but in his lifetime relatively few people can have read or heard it. He felt easy in the company of poets, getting to know Walter Scott and Lord Byron in later years, while Coleridge and Southey duly admired his poetry.[31] He saw himself as 'poet, philosopher, and sage'; aspiring, in the new Romantic world, to be a Renaissance man. Unlike Darwin, he did not attempt to get science across in poetic form: he had read for Wordsworth the proofs of the second (London) edition of *Lyrical Ballads*, putting in the punctuation, and the new fashion was for a poetry of feeling rather than exposition or matter-of-fact instruction.

Although Coleridge said that he went to Davy's lectures to improve his stock of metaphors, it is harder to find chemical echoes in his verse than in Herbert's. John Keats' medical training left marks in his writing, and Percy Bysshe Shelley, while a student, had been an enthusiast for chemistry. Again, there are in his poetry (and in Byron's 'Cain') allusions to science and to the new world-views it was opening up. But it was Tennyson's great poem, *In Memoriam*,[32] where he coined the phrase 'Nature red in tooth and claw', which brought to his readers the aeons of geological time invoked especially by Charles Lyell, and the evolutionary speculations in the sensational pre-Darwinian work, *Vestiges*

(1844).[33] T.H. Huxley loved Tennyson's poetry, saw him as 'the only modern poet, in fact I think the only poet since the time of Lucretius, who has taken the trouble to understand the work and tendency of the men of science', and ensured that the Royal Society was represented at his funeral.[34] But while, on the other hand, there have been since 1800 a fair number of scientist–poets, they have not popularised science in a straightforward way: rather, they have expressed feelings and beliefs, or been wry and jokey about science and its practitioners, letting us into their world. Light scientific verse was a definite Victorian genre, making scientific activity more personal and less forbidding, and puncturing the pomposity of some men of science.

Parody could play an important part there, as it had with 'The Loves of the Triangles':[35]

> But chief, thou NURSE of the DIDACTIC Muse,
> Divine NONSENSIA, all thy Soul infuse;
> The charms of *Secants* and of *Tangents* tell,
> How Loves and Graces in an *Angle* dwell;
> How slow progressive *Points* protract the *Line*,
> As pendant Spiders spin the filmy twine;
> How lengthened *Lines*, impetuous sweeping round,
> Spread the wide *Plane*, and mark its circling bound:
> How *Planes*, their substance with their motion grown,
> Form the huge *Cube*, the *Cylinder*, and *Cone*.

The poem goes on to describe the smoke jack for roasting meat, and recount the deeds of Little Jack Horner and of Cinderella, ending the section with:

> Alas! That partial Science should approve
> The sly RECTANGLE'S too licentious love!

It would be fair to say that the image of science that readers of this journal would have acquired would have been negative – it was good for a laugh, in so far as it was not dangerously left-wing. This was not the image that Sir Joseph Banks and his associates in the Royal Society would want at all: from the viewpoint of distinguished practitioners, not all publicity is good publicity where the popularisation of science is concerned.

Another anti-French publication of 1799 made the political point about 'philosophy', which included what we would call 'science', more explicitly, though less elegantly:[36]

> Philosophy, of Gallic climes,
> Parent of unexampled crimes!
> Philosophy who, while the clouds
> Bright Revelation's day, unshrouds
> DARK-LANTERN OF REGENERATION,

> That Will-o'-Wisp of the GREAT Nation,
> Whose glimmering sparkles emanate
> From Rotten Pediment o' the State,
> Just as stale fish and carrion trash is
> Known to emit electric flashes.

By the time Darwin's *Temple of Nature* came out in 1803, to a cool public and critical reception, Davy was promoting pure and applied science at the Royal Institution; and the poetry he genially parodied was his friend Wordsworth's[37] – some stanzas go:

> My cousin was a simple man
> A simple man was He
> His face was of the hue of tan
> And sparkling was his eye –
>
> He then became a farmer true
> And took to him for aid
> A wench who though her eye was blue
> Was yet a virgin maid.
>
> He married her and had a son
> Who died in early times
> As in the churchyard is made known
> By poet Wordsworths Rymes.

Though he wrote some light verses as thank-you letters, Davy normally used poetry for serious reflections, about nature, life and death – a natural philosopher's world-view. He envied the ancient Roman pagan his enchanted world-view:[38]

> For thee, the Eternal Majesty of heaven
> In all things lived and moved, – and to its power
> And attributes poetic fancy gave
> The forms of human beauty, strength, and grace.

The immortality of the soul, and a delight in the natural world verging upon pantheism, feature strongly in his verse, as they do in his *Consolations in Travel*,[39] most notably in that pantheistic rhapsody when he presented himself as poet, philosopher and sage.[40] In *Consolations*, written like Boethius' *Consolations of Philosophy* while facing death, he indeed appears (in the guise of 'The Unknown') as a sage, the role which was being played by his friend Coleridge (and was to be by Thomas Carlyle, Matthew Arnold and John Ruskin) – but he had got there through science ('philosophy') as well as poetry. There were not yet two or more cultures; but one of the problems facing natural philosophers, in an era when Baconian empiricism was fashionable, was that science could seem banal and utilitarian – a source of information and devices rather than of wisdom. Notably those associated with the Royal Institution, Davy, Faraday, Tyndall and Huxley, all sought to show in their lectures and writings, and in Davy's case also in verse, that this was not so: science was momentous, it required boldness and controlled imagination, it was the sphere of reason and not just understanding. This was and is not always easy to get across to the public.

Davy wrote a poem about eagles, whose soaring up into the Sun could symbolise the role of the scientist, and who taught their young to go higher by example:[41]

> The mighty birds still upward rose,
> In slow but constant and most steady flight,
> The young ones following; and they would pause,
> As if to teach them how to bear the light,
> And keep the solar glory full in sight.
> So went they on till, from excess of pain,
> I could no longer bear the scorching rays;
> And when I looked again they were not seen,
> Lost in the brightness of the solar blaze.
> Their memory left a type, and a desire;
> So should I wish towards the light to rise,
> Instructing younger spirits to aspire
> Where I could never reach amidst the skies,
> And joy to see them lifted higher,
> Seeking the light of purest glory's prize.
> So would I look on splendour's brightest day
> With an undazzled eye, and steadily
> Soar upward full in the immortal ray,
> Through the blue depths of the unbounded sky,
> Portraying wisdom's boundless purity.
> Before me still a lingering ray appears,
> But broken and prismatic, seen thro' tears,
> The light of joy and immortality.

Davy wrote this in 1821 on holiday in the Highlands, when he had recently been chosen as President of the Royal Society, and was recognising that his days of blue skies research were behind him. Many scientific lives climax, intellectually speaking, rather young – but the promotion, organisation and administration of science are an essential part of the public and practical activity we call science.[42] Unfortunately, Davy's relations with his most important protégé, Faraday, were not as happy as the eagles with their young: he failed to realise in time, in his role as a 'father in science', that his 'son' was grown up and needed independence, and a mighty row ensued, as can happen in the best families.[43] We now have the PhD degree, which corresponds to the end of apprenticeship; but the shadow of their scientific parent, above them in the sunshine, can lie heavily upon scientists, perhaps especially if the 'parent' (the 'doctor–father') was of the very first rank, and thus carrying enormous authority.

Lavoisier's lucrative post in the tax administration gave him an accountant's eye view of chemistry, where the books or equations must balance: in particular, the weights of the products of a chemical change must be the same as that of the reactants. This perception was the key for dismissing the existence of phlogiston, which appeared to have negative weight: matter was indestructible, mass conserved. Davy turned this law to poetic effect about 1818, incorporating his concerns about death:[44]

> The massy pillars of the earth
> The inert rocks, the solid stones,
> Which give no power, no motion birth,
> Which are to Nature lifeless bones,
>
> Change slowly; but their dust remains,
> And every atom, measured, weigh'd,
> Is whirl'd by blasts along the plains,
> Or in the fertile furrow laid.
>
> The drops that from the transient shower
> Fall in the noonday bright and clear,
> Or kindle beauty in the flower,
> Or waken freshness in the air:
>
> Nothing is lost; the ethereal fire,
> Which from the farthest star descends,
> Through the immensity of space
> Its course by worlds attracted bends,
>
> To reach the earth; the eternal laws
> Preserve one glorious wise design;
> Order amidst confusion flows,
> And all the system is divine.
>
> If *matter* cannot be destroy'd,
> The *living mind* can never die;
> If e'en creative when alloy'd
> How sure its immortality!

Later, in 1825, as his health began to give way, he elaborated a little:

> So may we hope the undying spirit
> In quitting its decaying form
> Breaks forth new glory to inherit,
> As lightning from the gloomy storm.

But clearly this poem has little to do with getting science across. Poetry like this gives us insight into the minds of scientists, rather than information about the world.

Davy's scientific papers were eloquently written, for a much less-specialised readership than scientific journals would find later. He resisted the theory-based language that had given us 'oxygen' and called the greenish suffocating gas he recognised to be a chemical element 'chlorine', from the Greek word for its colour. But he had earlier decided that the extraordinary substance, comparable to the alkahest, that he had isolated electrically from potash, was metallic: this despite its being so light that it floated upon water, decomposing it so violently that the released hydrogen caught fire. Having started with the term 'potagen', after consulting with colleagues and confirming its metallic if anomalous status, he switched to 'potassium', and subsequently 'sodium' for its congener. The convention is that metals' names end in 'um'. Davy had sufficient classical training in his year at Truro Grammar School[45] not to be at a loss in coining names from ancient languages. Margaret Thatcher, chemist turned politician, coined the phrase 'the oxygen of publicity', but on the whole words from modern science work chiefly in comic verse: we would be lost without 'catalyst' and 'chain reaction', but they are used vaguely and are hard to get into serious poetry. When Faraday, whose formal education had been elementary, needed new words, he consulted William Whewell, the Cambridge polymath and pundit, and together they came up with 'electrode', 'ion' and other terms, carefully defined and essential in electrical science but not playing much part in popular poetry.[46]

Whewell had been involved in the coup that jerked Cambridge mathematics into the nineteenth century, with Charles Babbage, George Peacock and John Herschel. They translated an up-to-date French textbook, and when appointed examiners, announced that they would set questions involving Leibniz's notation for the differential and integral calculus rather than Newton's, and with it modern French mathematical analysis. Babbage, the pioneer of computing, occupied Newton's professorship and was a keen populariser especially of technology; and Herschel, the son of the discoverer of the planet Uranus, became the great scientific sage of the early Victorian period, and was appointed, as Newton had been, Master of the Royal Mint. His *Preliminary Discourse*[47] was written for a popular series, Dionysius Lardner's *Cabinet Cyclopedia,* and became a classic in philosophy of science, admired by Whewell, John Stuart Mill and Charles Darwin. He wrote other accessible works,[48] but took a firm line on the place of amateurs in developed sciences, in this case astronomy, in an accompanying volume:[49]

> Admission to its sanctuary, and to the privileges and feelings of a votary, is only to be gained by one means, – *a sound and sufficient knowledge of mathematics, the great instrument of all exact enquiry, without which no man can ever make such advances in this or any other of the higher departments of science, as can entitle him to form an independent opinion on any subject of discussion within their range.*

Herschel's photographic portrait by Margaret Cameron, showing him as an old man with flying white hair, has become famous as an icon of a scientific hero; but Herschel was not only a physicist who refused to specialise[50] and became a sage,[51] but also a poet. He wrote, as Victorians were wont to do, a magnificent poem in classical metre (tricky to read, with a not-always-obvious caesura in each line), on the Baconian theme, 'Man the Interpreter of Nature', where the late appearance of mankind, of appreciative mind and scientific bent, in the geological timescale is greeted:[52]

> Man sprang forth at the final behest. His intelligent worship
> Filled up the void that was left. Nature at last had a soul.

We may recall that Herschel's friend Oersted[53] published a collection of lectures, essays and dialogues under the title *The Soul in Nature*.[54] He did not include verses, but at that time natural philosophers were feeling the need to express themselves and make it clear that they were no nerds: poetry was a vehicle for that.

In another poem, written for Peacock, by now ordained, made Dean of Ely Cathedral and preaching there, Herschel alluded to Peacock's innovations in algebra as a mace, the only weapon medieval clergy were allowed – being a club it did not draw (much) blood:[55]

> The organ's swell was hushed, – but soft and low
> An echo more than music rang, – where he,
> The doubly-gifted, poured forth whisperingly,
> Highwrought and rich, his heart's exuberant flow,
> Beneath that vast and vaulted canopy.
> Plunging anon into the fathomless sea
> Of thought, he dived where rarer treasures grow,
> Gems of an unsunned warmth, and deeper glow.
>
> Oh! born for either sphere, whose soul can thrill
> With all that Poësy has soft or bright,
> Or wield the sceptre of the sage at will,
> (That mighty mace which bursts its way to light)
> Soar as thou wilt, or plunge – thy ardent mind
> Darts on – but cannot leave our love behind.

Peacock had two vocations, as mathematician and minister; Herschel saw science as a calling, coming (despite his orthodoxy) close to Davy's nature-worship:[56]

> To thee, fair Science, long and dearly loved,
> Hath been of old my open homage paid;
> Nor false nor recreant have I ever proved,

> Nor grudged the gift upon thy altar laid.
> And if from thy clear path my foot have strayed,
> Truant awhile, – 'twas but to turn, with warm
> And cheerful haste; while thou didst not upbraid,
> Nor change thy guise, nor veil thy beauteous form,
> But welcomedst back my heart with every wonted charm.

Herschel's verse, whether original like these or translated from the German, was serious; but later generations saw the comic possibilities of scientific verse, if not perhaps in the style of the *Anti-Jacobin*. By the mid-century, the technical language of science was becoming more compressed; there were more people involved, and editors of scientific journals began to demand a more condensed style, addressed to fellow-experts who did not need much preamble. Lavoisier's ideal of an unambiguous elite language was being realised, and chemistry and physics were increasingly presented in equations and tables rather than prose (or verse). This recondite language could be used, especially among scientists (though that word was hardly yet coming into use) effectively in light verse, perhaps making serious points.

Thus James Clerk Maxwell,[57] one of the greatest of physicists, wrote verse which his first biographers[58] classified into 'juvenile', 'occasional' and 'serio-comic'. The occasional ones reveal his deep religious feelings, and his love for his family home in Scotland; but it is the last set which gives the public the clearest picture of a scientist being playful. Thus (as in the 'Loves of the Triangles') we have 'a problem in dynamics' expounded in verse, but this one refers to a whole series of genuine equations given at the end. We meet dynamics lightly in another poem, in dialect:

> Gin a body meet a body
> Altogether free,
> How they travel afterwards
> We do not always see.
> Ilka problem has its method
> By analytics high;
> For me, I kenna na ane o' them,
> But what the waur am I?

In parody of Tennyson, we have a lecture to a female student on electrical measurement:

> The lamp-light falls on blackened walls,
> And streams through narrow perforations,
> The long beam trails o'er pasteboard scales,
> With slow-decaying oscillations.
> Flow, current, flow, set the quick light-spot flying,
> Flow current, answer light-spot, flashing, quivering, dying.

But Tyndall, Faraday's successor at the Royal Institution, a self-made Irishman, looked at askance by Cambridge men, set Maxwell's teeth on edge with his presidential rhetoric at the Belfast meeting of the British Association in 1874, his wildly enthusiastic embrace of both scientific naturalism and imagination,[59] and his admiration for Germany (and hence tendency to give credit in thermodynamics to Germans rather than Scots). Maxwell mocked him in various verses, but especially in a 'Tyndallic Ode' which contains stanzas mocking Tyndall's style, and the content of a lecture on flames:

> I come from empyrean fires –
> From microscopic spaces,
> Where molecules with fierce desires,
> Shiver in hot embraces.
> The atoms clash, the spectra flash,
> Projected on the screen,
> The double D, magnesian b,
> And Thallium's living green.
>
> I light this sympathetic flame,
> My faintest wish that answers,
> I sing, it sweetly sings the same,
> It dances with the dancers.
> I shout, I whistle, clap my hands,
> And stamp upon the platform,
> The flame responds to my commands,
> In this form and in that form.

Thallium had been discovered using the spectroscope by William Crookes, another self-made London man of science. The second stanza perhaps brings vividly to life Tyndall's performances on the rostrum: though he could write accessibly, Maxwell was no great lecturer, and may have been somewhat envious. The period was, however, already coming when too great success with the general public might no longer enhance a reputation among men of science. Gentlemen of science frowned upon those who courted publicity; they were no longer one of us.

There are modern scientists who write verse, often light, and there were in the Victorian era. The Red Lions were a group that met socially at British Association meetings, taking their name from the pub in Birmingham where they had first gathered. When, in 1869, the B.A.A.S. was to meet in Exeter (where a new museum had been built as part of the welcome) 'Snug the Joiner' (who played the lion in *Midsummer Night's Dream*[60]) published *Exeter Change* in preparation.[61] The title is a pun: Exeter Change in London had been the site of a menagerie. On the reverse of the title-page is an advertisement for the forthcoming new journal, *Nature*, aiming first to inform the public, and second, scientific men outside their specialisms. The rest is less serious. There are spoof advertisements (amid a few

60   *Poetry, metaphor and algebra*

real ones), such as for a chain of reasoning lost by Professor Benjamin Brodie at the Chemical Society,[62] reports of fictional meetings involving scientists prominent in the B.A.A.S. and poems – this one, a parody of Tennyson, by a Mathematician (member of Section A), and set in Exeter:

> Tomkins, leave me here a little, while the section work is on:
> Leave me here to write a letter; I shall bless you when you're gone.
>
> 'Tis the place where I, astounded, read the notice of my doom
> Dreary seems the morning paper in this dull Reception Room.
>
> Many a night, at former meetings, have I, as I went to rest,
> Thought I could my Kate rely on, hoping wholly for the best.
>
> Many a night I led her safely through a crush of members famed,
> Brought her ices, cakes and coffee, told her how the things were named.
>
> Once about the beach we wandered, spooning in a way sublime,
> While the dreary tales of Science were reserved for future time;
>
> When geologists before us on contorted strata prosed,
> We were sure to find a seaweed in a place not much exposed.

Alas she has picked a husband from Section D (Zoology and Botany) instead!

> But what matters! I'll forget her: seek another in her place:
> Some downright strong-minded woman, fit to rear my musty race.
>
> Iron-visaged, tall and skinny, she the bluest of the blues,
> Shall with potent mathematics my domestic life suffuse.

But the story ends in flight:

> But howsoever this may be, a sad farewell to this old place:
> Not for me are nice excursions where I might my cousin face.
>
> No, I'll seek the Queen-street Station, for the time-bills plainly show
> I may catch a train inviting, running homeward, and I go.

Other poems are less-accomplished, and on less-promising topics, including the British Pharmaceutical Conference, meeting with the B.A.A.S.; the *Eozöon Canadense*, a primitive organism supposedly found in slime; and the history of chemistry, with a stanza about Lavoisier, recently hailed by the patriotic Adolphe Wurtz[63] (trying to get the French government to invest in science) as the founder of chemistry, of immortal fame:

> O Lavoisier, master great,
> We mourn your awful fate,

But never tire of singing to your praise;
And if what Wurtz says be true,
We need but trace to you
The chemistry of our enlightened days.

The verses in this publication were for those attending the meeting and, as is clear, bringing their families and making a holiday out of it. But the comic magazine *Punch*, with a wide general readership, also saw features of science worth making a song and dance, or anyway a poem, about.[64] Darwinian evolution, and especially our kinship with the apes, stimulated cartoonists and versifiers alike;[65] and so did the foul state, and subsequent improvement by sewage works, of the Thames:[66]

The passenger of Chelsea boat
Unwonted salmon shall admire,
Where dogs and cats he used [to] note,
Defunct that on thy breast did float,
Emitting exhalations dire.

While elsewhere, stimulated by an almanac, *Punch*'s team had fun with 'Mrs Durden', a member of the public with little sympathy for or understanding of science:

Your monsters with them crackjaw names described by LYALLS and JOE MILLERS,
What if they was but dragons slain by early saints and giant-killers?
And how if somebody, by'nd by, beyond a doubt succeeds in provin'
As how the earth is standin' still, the sun is, as 'e looks, a movin'?

Your scientific wonders is, in my opinion, the invention
Of one whose name it isn't thought polite and proper for to mention.
I thinks when he found out that folks in witches had got unbelievin',
That he put into wizards' heads them other methods of deceiving.

The old wife looked forward to the good old times coming back, when the 'delugions' of gas, electricity or steam would be banished 'on a suddent'.

Punch was at least as famous for its illustrations, cartoons which in Victorian times usually had several lines of caption, as for its verses; and Darwin with his beetling brows was a wonderful subject for the caricaturist. In alchemy and botany, where we began, visual language was of extreme importance; and just as some of the most exciting and attractive Victorian buildings were designed by engineers rather than architects, so scientific illustrations of places, plants, animals, machinery, apparatus and processes have aesthetic qualities more striking than the sentimental genre paintings produced by professional artists and beloved of our ancestors. So we turn to pictures.

5   Picturing science

We all remember Alice's questioning the use of a book without pictures or conversations. Conversations were indeed a way of getting people into science: Jane Marcet's *Conversations on Chemistry* were a classic of this genre, as were Jeremiah Joyce's *Scientific Dialogues*. No doubt others beside the young John Stuart Mill enjoyed this little book (whatever stern and ambitious fathers thought about it), and other schoolchildren learned it by rote, like the uncomprehending pupils Bishop Heber found in India.[1] It did indeed make science more attractive than blocks of prose would have done. Both these books had pictures too: Joyce's title page depicts the Leaning Tower of Pisa, to make a point about Galileo, while opposite is a frontispiece of a bewigged dominie conducting a lesson for three boys (though in some of the dialogues an Emma appears). Jane Marcet's more significant work had an all-female cast, with a governess and two bright charges.[2] There, the illustrations are grander, engravings occupying a whole page, and show apparatus and how to use it – disembodied hands hold pieces the right way round. While the chemistry she got across was quite austere (no gendered concessions to cookery or needlework), the text's attractiveness was undoubtedly increased by the pictures.

Illustration was a feature of chemical books,[3] because chemistry was (and still to a great extent is) essentially an experimental science. Manipulation is crucial. Careful handling and exact weighing and measuring are necessary skills; and while they can only be properly learned on the job, good pictures are enormously helpful. Michael Faraday's, in his classic 'how to do it' book, *Chemical Manipulation*,[4] were woodcuts set in the text, less expensive to do than Jane Marcet's (or Lavoisier's) copperplate engravings, and also more convenient because the reader does not have to keep one finger by the illustration while holding the book open with the others. Chemical illustration has a long history, with more to depict than just apparatus: but it is also a development, not unlike that with written and spoken scientific language, from decorative wealth to austerity and clarity. The aesthetic richness of pictures of the early days of the science gives way by the beginning of the twentieth century to diagrams, intelligible only to those already experienced in chemistry. This evolution we shall follow,[5] and then look at other sciences, physics and astronomy, and particularly natural history, to see whether the same pattern emerges; and then look at how illustration has both made sciences attractive, and also conveyed messages that cannot be got across even in a thousand words.[6]

Chemistry has always been closely concerned with a visual language of illustrations and symbols. To outsiders, chemistry has often been arcane, and looked as if it were intended to baffle them – indeed to blind them with science. In alchemy, pictures of kings, snakes and lions, the Sun and the Moon, were never (as in any interesting art) straightforward to read: like the texts (often in verse) that they accompanied, they were ambiguous, operating at different levels and deeply intelligible only to the adept, who had spent years working with a master. There was no one right way to understand them: they were not simply coded recipes, and there was no single right way to illustrate the chemical process.[7] Symbols were similarly uncertain. Metals and planets, and thus ancient gods like Mars (iron) and Saturn (lead), were identified – but not always consistently. Not only was the course and outcome of reactions uncertain, given the impurity (in our terms) of the reagents, but different authors used different symbols to represent substances, or were not focused upon what we would see as a single substance.

And just as calligraphy in the scripts of languages we do not understand can be very attractive, and religious paintings in alien or exotic traditions deeply moving, so we can find alchemical paintings beautiful and significant. Here as elsewhere, but even more so, meaning is fluid, and what people get out of pictures may not be at all what the author expected or put in. Alchemists, working in a genuine if occult tradition, but casting pearls before swine, would rejoice at not being understood by dilettantes like us who have not served an appropriate apprenticeship.[8] They were seeking to express hard-earned wisdom, the gold concealed in dross, the significant behind the ordinary. Most of their contemporaries they would see as superficial, men who look on glass and on it stay their eye. They would have had us pass through, like Alice going through the looking-glass, into the topsy-turvy real world.[9]

By 1789, when Lavoisier's *Elements of Chemistry* was published, chemists wanted to be understood. Joseph Priestley published careful accounts of his experiments on gases so that others would be able to repeat them. He got much of his apparatus specially made for him by Wedgwood, who subsequently made more for sale; and his eventual book was illustrated with nine handsome engraved plates, two of them folding.[10] Lavoisier was very rich and his laboratory at the Paris Arsenal was splendidly equipped; his *Elements*[11] had thirteen plates, some folding, and some with a scale to show the (impressive) size of some pieces. Priestley hoped that all his readers would join in the splendid and enjoyable practice of experimental chemistry, all adding a brick or two to the edifice of science. Lavoisier took a more elitist view, expecting science to be advanced by men of genius like him with excellent equipment; but he did hope that those who carefully repeated his experiments according to his instructions would get the same results and come to the same conclusions, notably about oxygen being the key to burning. Newton had sent a prism to someone who could not get the same result in an optical experiment: Lavoisier expected that good pictures would do the job.

In the magnificent portrait by Jacques Louis David of Lavoisier and his wife (in which she seems the dominant figure) now in New York, there are some pieces of apparatus beautifully painted at the great man's feet.[12] The engravings in his book are also carefully shaded with hatching to indicate the three-dimensional form of the vessels, the frameworks which support them, and even the material of which they are made. After all, there was an aesthetic convention that pictures should be realistic, and part of the engraver's job was to improve amateurish sketches into professional-looking plates: if something was spherical, it was not good enough to portray it as a circle.

During the nineteenth century, this requirement was relaxed. The invention of photography led to the replacement of 'realistic' pictures; which had previously often been made with the assistance of optical devices like William Hyde Wollaston's 'camera lucida'. Drawing ceased to be an accomplishment expected of gentlemen. Painters became less preoccupied with perspective. Like illustrators of painstaking topographical works, or good do-it-yourself books, chemical authors had carefully indicated the bricks that supported a heated gun-barrel, or the clamps that held a condenser in place on a retort stand. After all, the work might be read by tyros, who needed all the help they could get if they were to get good results: especially in pre-pyrex days when glass apparatus had to be thick enough to stand up to mechanical shocks, and yet thin enough not to crack when heated – and was sealed by favourite concoctions called 'lutes'. Real understanding in hands-on chemistry had to be achieved through the fingers, but they could be guided by good pictures.

Painters after the mid-century experimented, following Goethe[13] whose polemical book on optics was translated by the first Director of the National Gallery in London, in using colour to achieve depth, and ceased (now that there was photography) to be so concerned about detailed representation. Gradually, as more people were formally taught chemistry, it became less necessary to be so realistic in illustrating that too: elegant, spare, linear, two-dimensional pictures came in so that, by 1900, shading to indicate form seemed rather quaint and old-fashioned. This change in visual language parallels the change from the clear prose of Priestley, Lavoisier, Faraday or Davy to the crabbed and technical language of scientific papers, intelligible only to the initiated. Pictures had become diagrams. In mineralogy by the middle years of the century, illustrations of actual pieces of rock (often coloured) were replaced by elegant idealised drawings of crystal forms, accompanied by chemical formulae.[14] Geometry came into mainstream chemistry about the same time with the croquet-ball and wire models displayed by August Hofmann in a famous lecture at the Royal Institution.[15]

Newton had had an 'elaboratory' in Cambridge, but few people in the eighteenth century, or in the first half of the nineteenth, had access to a room set aside and equipped for chemistry. The kitchen, with its range and sink, was an obvious place to use – for example by Jacob Berzelius, tsar of chemistry in the 1810s and 1820s, whose formidable housekeeper Anna washed the dishes and test-tubes. In his native Sweden, rich in minerals and with an important metals industry, surveyors had begun taking 'portable laboratories' with them on their travels, so that they could do analyses on the spot and see if there was gold (or some more useful metal) in them thar hills.

There are illustrations and then advertisements for the elegantly boxed collections of apparatus that was available by Berzelius' time, in the same way that doctors on their rounds had chests filled with medicaments, ready to be dispensed to patients.[16] Platinum had been isolated as a powder, but about 1800, the London doctor William Hyde Wollaston made his fortune by inventing a way of making it malleable and coherent by hammering at high temperature: nobody who could afford gold wanted silvery jewellery, but as an inert metal, it was a godsend for chemists. A scrupulous experimenter, known as the Pope because his analyses were infallible, Wollaston also devised ways of working with much smaller quantities of reagents than Priestley and his contemporaries had used. Chemistry, as Davy put it, had changed by 1830:[17]

> An air pump, an electrical machine, a voltaic battery (all of which may be upon a small scale), a blow-pipe apparatus, a bellows and forge, a mercurial and water gas apparatus, cups and basins of platinum and glass, and the common reagents of chemistry, are what are required. All the implements absolutely necessary may be carried in a small trunk; and some of the best and most refined researches of modern chemists have been made by means of an apparatus which might with ease be contained in a small travelling carriage, and the expense of which is only a few pounds. The facility with which chemical inquiries are carried on, and the simplicity of the apparatus, offer additional reasons, to those I have already given, for the pursuit of this science. It is not injurious to the health; the modern chemist's ... processes may be carried on in the drawing room, and some of them are no less beautiful in appearance than satisfactory in their results.

The girls in Jane Marcet's *Conversations* perform experiments in an ordinary room, and in their ordinary clothes (burning the muslin with spilled acid on one occasion), and no doubt she imagined a portable laboratory brought out for the occasion.[18] The few schools teaching science also used them. Faraday illustrated one for the handsomely illustrated textbook by William Thomas Brande,[19] Davy's successor; he had used a portable laboratory (or 'travelling apparatus') when accompanying Davy on his continental tour in 1813–15 for their research on iodine,[20] and another of his is preserved in the Science Museum in London.

From the beginning, in 1799, there was a well-equipped laboratory in the basement of the Royal Institution, beneath the lecture theatre – both of which the Royal Society lacked. It is also illustrated in Brande's book, but it was many years before such facilities became at all common. Davy relished doing at least some of his research in public; and in the laboratory there were seats banked up for small audiences, who would have got a commentary from the great man on what he was up to, in a lively form of laboratory instruction. But most teaching was, as upstairs at the R.I., done with carefully planned demonstration-experiments before an audience, whether of (medical) students, as with Brande and in other medical schools, or of interested members of the public. At the Royal Institution, they always worked: most people who have watched them elsewhere must remember best those that didn't, making chemistry lessons more fun.

Thomas Thomson in Glasgow (whose standard textbook the older Mill admired) had got his students working in the laboratory in the 1820s, and so had others in Germany; but it was Justus Liebig's well-publicised school of chemistry at Giessen that made laboratory instruction central.[21] His example led to teaching laboratories in other universities, notably at University College, London, where there is an illustration from 1846 showing work in progress: the students are wearing aprons (white coats came much later) and there is a man (perhaps the professor) in a top hat; there are long top-lit benches with stools (as in a bar) upon which experiments are in progress, and above which bottles are ranged.[22] Important publics came to understand that practical instruction and experience in the laboratory was essential for learning chemistry, and later other sciences too.

By 1875, when a Royal Commission published its report on the teaching of science, a few 'public' schools had laboratories: there are plates of those at Rugby, Dulwich, University College School in London (which has banked seats like Davy's), and plans of those at Clifton, Harrow and Manchester Grammar School.[23] The Commission (chaired by the Duke of Devonshire) had been set up in 1870, as universal elementary education at last became compulsory in England. The Franco-Prussian War of that year, in which the better-educated nation won a stunning victory, was a great spur to the advancement of science. The message from the Commission's report was that all secondary schools should teach some science, and the illustrations of best practice were to spur them to do likewise. Again, pictures could get a message across clearly. They did; and, as late as the 1950s, teaching laboratories had hardly changed. The best of what we are apt to think of as the rather dull but profitable genre of textbooks were, like Brande's early example, effectively illustrated[24] – but showing what a laboratory looked like became less and less necessary.

The story is not very different in the realms of physics, astronomy and engineering. The great tome on 'mathematical instruments' by Nicholas Bion, instrument-maker to the King of France, was translated in 1723; the updated edition of 1758 contains thirty folio-size plates, mostly of apparatus.[25] By then, telescopes and surveying equipment were being made to high standards and were available as standard, 'off the peg'; gleaming with brasswork and polished steel, many are very handsome.[26] The tradition of fine craftsmanship continued into the nineteenth century,[27] and surviving pieces have become collectors' items. The young King George III had a splendid set of apparatus for experiments in natural philosophy (physics), now in the Science Museum in London.[28]

It may be that items made partly for show, like early globes and microscopes to grace gentlemen's libraries, or for demonstrations before opulent audiences, have survived better than workaday apparatus. These were often cannibalised, bits and pieces (perhaps especially lenses) being re-used under new circumstances: Davy creatively misused apparatus in his research; Faraday seems to have hated throwing things away, and recommended ways of recycling, making do and mending. Anything made of glass was liable to break, so that chemical ware has survived less well than objects of brass. Illustrations were important to contemporary users and to us, to indicate just what the instrument was and what it did.

Astronomical works were generally illustrated, but particularly noteworthy was John Pringle Nichol's *Architecture of the Heavens* (1838, 2nd edn, 1850). The second edition was dedicated to the Countess of Rosse, whose husband had built an enormous reflecting telescope in his native Ireland, six feet across: a marvel of precision engineering. The climate of Ireland does not favour astronomical observation, but Lord Rosse managed to resolve a number of nebulous objects in the sky into congeries of stars – a process Galileo had begun with the Milky Way two centuries earlier. Nichol illustrated on a black background the amazing sights visible through huge modern telescopes built since the Herschels had shifted the attention of astronomers beyond the solar system back into the region of the stars.[29]

Some of the nebulae in these spectacular plates were spiral in form, which might reinforce the 'nebular hypothesis' of William Herschel and Pierre-Simon Laplace that the solar system had developed out of nebulous matter set spinning – accounting for the fact that the planets' orbits all lie in a plane. But clearly many of the nebulae were composed of numerous stars so far away that they looked like a blur, and so had nothing to do with the origin of planets. Others might indeed be genuinely composed of hot gas, 'fire mist', but the nebular hypothesis, beloved of Robert Chambers in his (unillustrated) *Vestiges* as the beginning of the evolutionary process, remained controversial.[30]

Nichol was Professor of Practical Astronomy at Glasgow, and gave William Thomson, the future Lord Kelvin, an impulse towards physics in which he became a giant. Clearly, the title of the book (*Architecture of the Heavens*) indicates its commitment to a Divine Architect; and as well as astronomical plates, the book contains lithographed drawings by the Scottish artist David Scott who had 'consented to describe, by his extraordinary pencil, some of the emotions and aspirations produced in him by the loftier speculations of Astronomy'.[31] Sages gazing up into the heavens, entranced disciples, classical deities and nymphs in various stages of undress, are accompanied by suggestive quotations or phrases. The book ends with a section on 'Psyche or evolution', exploring universal evolution in the spirit of Erasmus Darwin: and the naked torso of Psyche is drawn against a ladder (not Jacob's, but something which looks more like a fire-escape) with the lines 'Audacious Psyche! Seek'st thou to ascend/Coveting the inaccessible?' The last picture is of a misty mass of bodies, captioned: 'Painful is birth/Painfuller life/Passing to that higher life the painfullest.' The plates, alternating with the superb astronomical ones, give a curiously troubling and mysterious aspect to astronomy, bringing Heaven and the heavens together, in an extraordinary piece of scientific communication.

68 *Picturing science*

Much more down to earth were engineering drawings. With the Industrial Revolution came illustrated patent specifications,[32] and the nineteenth century was the great age of the drawing office, and the expectation that plans would be followed exactly with nothing left any longer to the imagination or experience of the craftsman.[33] This applied to ships and buildings as well as to locomotives and bridges, and the drawings have an austere elegance which makes them sometimes, like the structures, works of fine as well as useful art. Colour washes may be used to indicate materials rather than ultimate appearance; and to interpret technical drawings is not and was not easy for the uninitiated. For the boardroom, therefore, handsome artist's impressions of the ship or locomotive in appropriate livery would also be prepared. We therefore have both the representational and the diagrammatic, complementary and both in their way showing what the machine is like. Something like this also happens, in a much more abstract way, in different representations of molecular structures in modern chemistry.[34]

Industry had been well-illustrated in the great *Encyclopédie* of Diderot and D'Alembert, which is often associated with Enlightenment irony about church and state, but which actually had much more about arts and crafts.[35] Later encyclopaedias, some of them explicitly technical, followed this tradition, and we find nineteenth-century machinery well pictured. Equally informative are the guides to international exhibitions, notably the Great Exhibition of 1851, where the Crystal Palace was itself the most striking exhibit. Thus the *Illustrated Exhibitor*, a handsome, large volume published in parts through 1851,[36] has a fold-out frontispiece over 60 cm wide of the palace surrounded by crowds of people and carriages; another plate, over 45 × 35 cm, shows Victoria and Albert enthroned within it amidst a great throng of guests, whilst others show areas of the building, and individual exhibits of machinery, manufactures and works of art. The profit from this exhibition (the first, and only one so far, to make one) went to buy land in South Kensington, where now the Natural History Museum, the Science Museum and the Victoria and Albert Museum stand: all these, in their way, carrying on, and illustrating, science (pure and applied) and design applied to technology.[37]

The Great Exhibition was concerned with fine art as well as with science and technology; and these two spheres came directly together in the many splendid volumes of plates of the construction of the railways, which made possible the building of the Crystal Palace, and the pilgrimages of millions to London to see it.[38] These volumes combined picturesque and topographical art with the careful depiction of tracks, stations, trains, passengers, signalmen, drivers, porters, navvies and rustics about their business. The most magnificent were the elephant folios of John Bourne, devoted to the construction first of the London and Birmingham railway.[39] This terminated at Euston: from there trains were hauled up by a stationary steam engine to Camden Town, where locomotives were attached. The enormous works associated with building a railway, essentially with picks and shovels, are vividly depicted: we see embankments, tunnels and bridges, some under construction, and some complete with trains running. As well as getting attractive views of Victorian countryside, one can see how the tracks were held in place, and how the points were changed.

Bourne then described the Great Western railway, from London to Bristol, built by Isambard Brunel to a broad gauge of seven feet, instead of the standard four foot eight-and-a half inches which is supposed to go back to Roman vehicles. Everything about that railway was on a grander scale; and, in both books, there were commentaries upon the scenery, geology and antiquities of the places the line passed through. London was not the first place to get railways, which originated in the north,[40] and books of rather less but increasing magnificence, but similar character, had been published on the lines between Liverpool and Manchester, Newcastle and Carlisle, and Manchester and Leeds.[41] This last one included a timetable, and was published by Bradshaw, whose name came to stand for the national timetable as the various lines grew and coalesced into a system.

All this indicates an admiration for new technology, which in fact went with the romantic world-view (especially but not only the sublime) better than one might have expected:[42] we also meet the industrial picturesque in studies of coal mines.[43] Most of these splendid books were themselves illustrated by a new technique: lithography. Copper plates had replaced woodcuts in the mid-seventeenth century, because they could be much more accurately detailed, but were very expensive: between five and twenty guineas (£1.05) in the early nineteenth century for each plate worked by a skilled craftsman.[44] This added greatly to the cost of books. The lines were engraved, pushing a burin; or scratched through varnish and then etched by acid, giving a less hard line. Plates were printed in a different press, and on a slightly different sort of paper from that used for letterpress printing. Because copperplate is an intaglio process (what is to show black is cut down into the metal), while letterpress is a relief process (what is to be black stands up), the two cannot straightforwardly be combined on one sheet of paper: so the illustrations had to be more or less separated from the text they illuminated.

Lithography, invented at the very end of the eighteenth century, was much cheaper than engraving, and the artist might even be his or her own lithographer: the line is much more flowing than is possible in engraving, and by about 1830 similarly fine detail could be shown. For fine works, lithography became the norm, and in geology especially, the scope this gave for more pictures boosted the science greatly. The succession of strata, and the details of fossils, was much easier to draw than to describe.[45] If required, lithographs (like engravings) were hand-coloured using as a guide pattern-plates prepared by the artist, who had used a printed proof which thereby became the nearest there is to the 'original'.

Also at the end of the eighteenth century, Thomas Bewick in Newcastle perfected the technique of engraving on the very close end-grain of boxwood, inaugurating another method of illustration.[46] The box is a small tree, and early examples were all quite small, but highly detailed so that they can be effectively blown up. Later, different blocks were bolted together. This is a relief system and therefore the blocks can be set with the letterpress, and the illustrations appear in the midst of the text. The blocks were very strong, and could take longer print-runs than copperplates or lithographs: when stereotyping came in,[47] along with electrotypes,[48] revolutionising the printing industry, the pictures were included as 'clichés' for endless repetition. Metal engraving, but on steel instead of copper, also became important in Victorian publishing, but less for science than woodblocks or lithographs.

It was natural history that benefited especially from new methods of illustration.[49] Audubon's double-elephant folios were indeed engraved on copper at enormous expense,[50] but a little later the illustrations that John Gould assembled in his great ornithological works were lithographed.[51] The nineteenth century was a great age of illustrated zoology (especially birds) and botany.[52] One of the pioneers of lithography was William Swainson,[53] whose pictures of birds and shells, in octavo format, are superb. He had travelled to Brazil, where he was unable to get into the interior because of unsettled political conditions, but managed to assemble a collection of specimens. He hoped to get a job at the British Museum curating zoology, but it went to someone else; so (henceforward a rather awkward and disappointed man), he made his living by writing and illustrating.

He brought out his illustrated natural history in parts, so that the costs to subscribers would seem less: when the whole was complete, buyers could get the plates and brief descriptions sorted out into natural groups, and bound up into volumes. This system had the advantage that cash-flow problems should have been avoided, because the receipts for part one would pay for the production of part two and so on; but subscribers fell by the wayside, defaulted or died, and others who joined in later wanted the earlier parts to make up their set, and life was not therefore simple and easy for Swainson and others (including Audubon) who published in parts.

Swainson was also involved in less-expensive illustrated works. He wrote a volume on birds, flycatchers, for Sir William Jardine's *Naturalist's Library*,[54] one of a number of cheap series appearing in the much larger market for books.[55] The publisher, Lizars, began Audubon's immense work, but had been unable to carry it through – Jardine's series are small octavos. Swainson's delicate drawings were engraved and hand coloured, and the birds are shown against a lightly sketched-in naturalistic background: sixty years later, Alfred Newton remarked in his *Dictionary of Birds* (1893–6): 'These figures were drawn by that admirable ornithological delineator, and most of them for truth of detail or beauty of design have seldom been equalled and rarely surpassed.'[56] Swainson was also closely involved in a similar venture published by Longman, Dionysius Lardner's *Cabinet Cyclopedia* – 'cabinet' being a book-size suitable for the pocket, just like Jardine's small octavo. Unlike an encyclopaedia, these were distinct volumes written to stand alone, forming a wide-ranging and fairly stiff non-fiction series.

Lardner, notorious for declaring that steam ships could never cross the Atlantic, and for eloping to Paris with an army officer's wife, was supposed to have changed his name from the plebeian Dennis to autocratic ancient Greek, and was therefore mocked as 'the tyrant' by his team of authors – certainly his correspondence with Swainson, preserved at the Linnean Society, is full of unrealistic deadlines on one side, and implausible promises on the other. The volumes came out cloth-bound with paper labels: some traditional-minded buyers had them rebound in leather. Swainson wrote those on natural history, which were illustrated with little wood engravings scattered about the text as appropriate. Some, in the volume on birds for example, are of the whole creature – vultures perched upon mountain tops, the secretary bird in the desert[57] – but most are of details (claws, bills, feathers) valuable in classification, but aiming at usefulness rather than beauty.

Anybody writing a series of books on natural history, especially against the clock, needs some way of organising them. Swainson became an enthusiast for a system in accord with his Trinitarian beliefs and his conviction that nature must have a pattern:[58] the so-called 'quinary system'. Here, animals were arranged in groups which formed patterns of three circles, the lower one containing three smaller circles: the system was Trinitarian, but quinary (five-fold) in that there were on every level of the system two big circles and three small ones. This system interested the great evolutionists Charles Darwin, Thomas Henry Huxley and Alfred Russel Wallace, searching for a pattern: but they, and their contemporaries, came to see it as a strait-jacket imposed on nature rather than the way the world was. Nevertheless, the diagrams (like the shrubby evolutionary tree in the *Origin of Species*) are an important part of the illustration of the book: in natural history as in chemistry, there are various ways of conveying knowledge in visual language, with a gradation from the delightful portraits of flycatchers through dissections to abstract circles or hexagons, accompanied by Latin or other jaw-cracking technical terms.

It is a matter of some contention whether scientific pictures of animals, or medical illustrations for that matter, ought to be portraits of an individual or should be somehow generalised, so that they show a species of parrot or a victim of some disease. Edward Lear, famous for his limericks, made his reputation as a zoological artist. His most famous work in that line was his sumptuous *Parrots of 1832*,[59] done in hand-coloured lithographs. They were painted from life, in the recently-opened London Zoo, and they are astonishingly life-like compared to most illustrations of exotic birds, which were done from the stuffed skins of dead birds rubbed with arsenic to preserve them and sent back from abroad. Better than that were the illustrations done in India, by artists trained to do Mughal miniatures, of birds[60] and plants;[61] in China, of animals and birds;[62] and in Colombia, of plants;[63] all retaining subtly different features from European artists' work. Field naturalists had the maxim 'What's hit is history, but what's missed is mystery', and (before binoculars) were quick on the draw. The only way to be sure just what one had been watching was to kill it. But each individual victim has idiosyncrasies, and the zoologist or medical student needs to know about the average (or maybe the 'essential') parrot or human: some of Lear's specimens were untypical. Like other stay-at-home naturalists, working in a zoo or a museum, he could see few examples of any one kind of creature.

Animals and birds are hard to spot in their environment, being camouflaged:[64] they must be depicted in a more open situation than they are generally found, or perhaps, as with Swainson and Lear, the background may be lightly sketched. It may also be sensible to depict many more kinds of animal in a plate (or museum diorama) than one would normally see together, either to show species that inhabit the same region, or creatures that look like one another for comparison. The artist has to intuit or be instructed how to bring out what is important in classification rather than what is adventitious: like any other observational science, illustrating natural history is not simply a matter of having an innocent eye and open mind, but of being aware of the need to make sense of knowledge.

Through Audubon, the birds of America became known. His giant folios were exceedingly expensive and sold in small editions, but later versions were more accessible. In Australia, there were some expert engravers, exiled there for forging banknotes and other offences,[65] and from the beginnings of settlement there was a tradition of support for science, especially natural history, and a market back in Europe for illustrated works. But it was John Gould whose beautifully illustrated volumes did for Australia what Audubon had done for the USA.[66] He had begun them in England, but soon found it necessary to go to Australia for an extended stay to do the job properly.

The sexes of birds sometimes look very different, and Gould reported from Australia that occasionally they had been seen as different species. He and his contemporaries therefore sought to put male and female into their plates; and might include a nest, and perhaps prey or food, in a more or less naturalistic plate. Others showed the bird on a studio stump, but with an egg; sometimes indeed naturalists depicted the entire lifecycle of an organism – for butterflies, this meant eggs, caterpillars on a food plant, chrysalises, and adult males and females. The already busy plate might also be systematic, including very similar species for comparison; or ecological, with animals and plants found in the same neighbourhood. Botanical plates from before 1800 included dissections (indeed that distinguishes them from flower paintings) because the Linnean and later more natural systems depended on structures; and zoological plates also sometimes included dissections.

Geologists faced with incomplete fossil remains dotted in what they took to be broken off; and reconstructed extinct animals in what we could call a visual language of hypothesis. Miller, in his *Crinoidea* (1821), has a wonderful exploded drawing of a stalked starfish, or 'stone lily'.[67] Sometimes this guesswork was faulty: the iguanadon, the first land dinosaur to be found, was thought to crawl on all fours like a crocodile; but Huxley inferred from the hips of dinosaurs that many of them had walked on their hind legs like an ostrich, iguanadon being one. Despite the television programme, we cannot walk with dinosaurs; but landscapes carefully drawn on scientific expeditions, whether to the western USA to survey railroad routes, to central Australia, or into darkest Africa, often showed flora and fauna, including humans – 'natives' and explorers.

People perhaps wanted just a representative 'native', but humans are a kind of animal where even in an anthropological treatise we usually want to know what an individual looks like.[68] These categories include scientists: there is a splendid picture from about 1840 of a representative field naturalist at work searching a pond for insects; and portraits of William Buckland in his geologising costume, entering Kirkdale Cavern and embarking upon a glacier during his work on the Ice Age.[69] Buckland, who enjoyed playing the buffoon, was much caricatured, as were some of his contemporaries: notably, quack doctors were mocked, but also used pictures of themselves in advertising their remedies.[70] From spirited caricatures, we have a good idea of how people looked; but Buckland, and others from the upper ranks of science and society, also had their portraits painted. Often, as with Buckland and Davy, engravings from a portrait would feature as the frontispiece to a biography; or, as with Hugh Falconer, in the collected papers published late in life or after death.[71]

Falconer's portrait, though, is a photograph; and by 1868 we have a much better idea of what most scientists looked like because 'carte de visite' photographs, often sent to friends, acquaintances or disciples, had become common. Whereas painted portraits often gave an idea of what the sitter had done (Davy's safety lamp is in the background, Buckland holds a formidable hyena jawbone, Banks has a globe beside him), photographs were normally done in a studio with standard props, and scientists cannot be readily distinguished from civic dignitaries or businessmen. Photos are almost inevitably a likeness at a particular moment, a snapshot, in most cases revealing the exterior but not what lies beneath: which is what the great portrait painter, or indeed photographer, strives to bring out.

Caricature can do this, as with Buckland, and in the later nineteenth century cartoonists like 'Spy' aimed to catch the significant features of eminent people, including Huxley and Tyndall, distorting the image so that their heads are much too large, but indicating both stance (body language) and facial expression so that we imagine that we are seeing a speaking likeness. As Priestley had been in the 1790s, Charles Darwin became the great butt of cartoonists after the *Origin of Species* was published in 1859.[72] Here his famous heavy brows, more evident as he aged, were a gift – and especially helped when his face was attached to an ape's body to make a point about evolution in an economical and memorable image.

By the time Darwin died in 1882, the fuss had died down considerably; he was respected as a sage, and was buried in Westminster Abbey. It is this elderly Darwin, rather than the clean-shaven man who sailed on HMS *Beagle*, or the bewhiskered middle-aged author of the *Origin of Species*, who is familiar to us and has featured on bank notes. These earlier portraits, and those of contemporaries, are printed in the massive and splendid edition of his correspondence.[73] They give us a useful insight into Victorian fashions in facial hair, which one might have expected to get in the way of making careful observations. Margaret Cameron's celebrated photograph of Herschel, a troubled seer with white hair wildly windblown, is magnificent – an icon, alongside the familiar twentieth-century picture of Einstein, of a physicist forever voyaging in strange seas of thought, alone. It was not long (1877) before high-speed photography, in the hands of Edweard Muybridge, made it obvious to everyone how racehorses and people moved. The origins of the film industry are to be found in the solution of this zoological conundrum.

Sketches of the famous and infamous feature in physiognomical and phrenological treatises:[74] some of these feature natural philosophers, natural historians or medical men, but there is an extraordinary array of odd-shaped heads and funny faces. It would clearly be impossible to teach these sciences (which in the early nineteenth century seemed respectable if unproven) through words alone, without at least some china model heads marked out with where the bumps come. Illustrations work even better, because there can be so many of them: and in moving from the idealised and generalised china head (or its image on paper) to real heads (or to caricatures) there is the great advantage that the theory can be tested. If prominent churchmen lack the bump of veneration, philanthropists that of benevolence, men of science that of causality, and felons those of combativeness and destructiveness, then something is wrong about the structure of the science.

In the end these would-be sciences collapsed partly because of failure to fit empirically, but mainly as is usual because of lack of adequate theory: science has to be facts ordered, organised common sense, and requires structure as well as observations, and preferably repeatable experiments as well. Nature may be more important than nurture, but phrenologists did not prove it to be so. Statisticians also had a go.

A way to show agreements and differences visually is through charts and tables. These often, like equations, turn off readers who are not already devotees: but they can help in getting science across, especially when put in graphical form when they become a kind of illustration. Indeed, there have been wonderful examples of data presented in fresh and even startling ways, so that their significance becomes clear; and here again aesthetics plays an important role.[75] Graphical presentations are not merely a second-best for the innumerate: they communicate quantitative data efficiently, as natural history pictures communicate appearances and structures impossible to convey in sentences. Alexander von Humboldt, for example, included abstract data like isotherms in maps. Also, having climbed very high on what was believed the world's highest peak (Chimborazo in Ecuador) during his Latin-American travels, he drew what became a celebrated diagrammatic picture of an equatorial mountain showing vividly how the ascent took one through all the climatic zones. The top was like Spitzbergen: height as well as latitude determines weather.

Later, in Louis Agassiz's studies of glaciers,[76] the large and handsome lithographed views (in the 'geognost' tradition emphasising underlying structure[77]) have transparent overlays indicating and naming the significant aspects – which led him to recognise traces of ice-action all over northern Europe, inspiring Buckland. Thematic mapping became an excellent way of showing quantities such as death rates (in small-scale epidemics, or country-wide and international statistics), and trade.

Priestley invented the historical timeline; and block graphs showing the movement of prices, exports and imports, or death rates over time were in use by 1800, William Playfair being an important pioneer. The pioneering statistical work of another mathematician, Lambert Quetelet,[78] abounds in tables, making it rather heavy reading; but has at the back graphs (including bell curves) and thematic maps showing crime rates which bring the data to life. It was still, of course, possible to tell lies and damned lies using statistics; we must all have come across plausible-looking but specious charts (usually with distorted scales). But graphs in their various forms, including pie charts, had become by 1900 a very important way of communicating scientific information.

In chemistry, Lavoisier's algebra combined with geometry led to symbols, equations and diagrams which closed it off from the uninitiated; and in other sciences, visual style (like verbal) became more abstruse as there were specialists there to read it. The field guides and distribution maps of natural history put utility before aesthetics; and Herschel's point that, in physics, nobody without mathematical competence has a right to an opinion[79] became increasingly true. And yet we live in a strongly visual age, and in popularisation illustration has during the second half of the twentieth century become more and more striking. In an age of specialisms, we also all depend upon vulgarisation, though at different levels; the visual has never been unimportant, or a second best to algebra, but it is now as important as ever. This is so particularly in communicating enthusiasm for science, not only to attract the young into it, but to induce their elders to pay for it. That again is not altogether new: Davy and Faraday had faced the problem at the Royal Institution. But it is to ballyhoo that we next turn.

6 Ballyhoo

Such excessive claims are made for science, and there are so many 'breakthroughs', that many of us become cynical. This has been true ever since the early days of the 'Scientific Revolution', when great things were forecast by Francis Bacon, and by his disciples in the newly-founded Royal Society after the Restoration of Charles II in 1660. Knowledge was power. Successful kings had always known that about political power, but here were promises that the curse imposed upon Adam and Eve might be mitigated: labour and pain would be reduced, and death delayed. The vision of alchemists, those eternal optimists, was transformed: the gold would be earned through science-based agriculture and industry, with flourishing commerce in the new sphere of oceanic shipping. Bacon on his title-pages took over the arms of Columbus, showing the 'pillars of Hercules' at the entrance of the Mediterranean and a ship sailing through them, with 'plus ultra', further yet, as the motto. New worlds would indeed open up to those following the new method of cautious inductive science, aiming boldly at 'the effecting of all things possible'.[1] In his utopian *New Atlantis* [1627], Bacon described a prospering island run by an academy of sciences: Britain could play that role.

In the seventeenth century there was indeed great interest in trades on the part of the early Fellows of the Royal Society; gentlemen condescended to find out how things were done, no longer despising manual skills. 'How to do it' books were published,[2] so that readers could find out about the mysteries of carpentry or printing normally learned by apprentice craftsmen on the job. Crafts were also changing. Christopher Wren was one of the stars of the Royal Society,[3] and his new St Paul's Cathedral, built after the Fire of London and incorporating wonderful wood carving by Grinling Gibbons, was paid for by a tax on coal arriving in London from Newcastle. While the new sooty fuel was making the city increasingly dirty, coal mines along the Tyne and Wear were dug deeper, running into water that had to be pumped out. In the second decade of the eighteenth century, Thomas Newcomen invented a steam engine rocking a beam and driving a pump that could do the job. Power had hitherto come from people, animals, wind or water: this was a really new departure, in which heat led to mechanical work being done, steadily, day-in, day-out. The industrial world was arriving. Agriculture also, by the eighteenth century, came to look different, with increasing enclosure of common land, 'rack renting' and selective breeding moving from racehorses to sheep, pigs and cows.

Distinct from, and more practical than, the Royal Society, the Society of Arts was founded to circulate knowledge of improvements, and to offer premiums and medals for inventions and discoveries. Following the Dutch example, 'high farming', mixing animal husbandry with growing crops, was introduced. The manure stimulated growth, and other fertilisers such as lime and marl were also brought into wider use. Improved ploughshares, hardened on one side so that they remained sharp as they wore down, and horse-hoes to weed efficiently between the lines of seedlings sown by a seed-drill, were introduced; potatoes became common, and crops such as the Swedish turnip were fed to cattle during the winter. A leisured consumer society, putting value upon gentility, civility and culture, began to make its appearance in Georgian Britain; though, naturally, the vast majority hardly yet shared in it.

All this involved very little of the recondite mathematical and experimental science that Descartes, Galileo, Boyle and Newton had been so prominent in introducing. It was common sense, open-minded empiricism, the work of hard-headed people without serious training in mathematics or science. The same was true of the pioneers of textile manufacturing machinery, like the spinning jenny: an intuitive feeling for machinery, capacity for experiment, refusal to be daunted when things did not work out first time, were the key. 'Self-help' and 'character' were the qualities Samuel Smiles saw in his heroes: practical, self-made and unpolished, grittily determined, but not 'applied scientists' or 'technologists' – such labels only make sense much later.

Eighteenth-century Oxford and Cambridge taught classics and some mathematics rather gently, until their syllabuses were reformed at the turn of the century. There were, indeed, professors of theology as well as medicine, chemistry, botany and other more modern disciplines, but their lectures were optional and their subjects no part of the regular curriculum. Dissenters from the Church of England, unable to attend these universities if conscientious, went instead to 'Dissenting Academies' where modern languages, natural philosophy and useful mathematics were usually taught, in courses originally intended to educate ministers, but attracting others headed for professional or business careers. They might also go to Edinburgh, with its flourishing medical school and lack of religious tests; to Glasgow, where medicine and science were also available; or to the Netherlands. Göttingen in Hanover, where the Elector was also King of England, was a new university much more modern in its courses than most, and attracted some British students more interested in what they were learning than in social cachet. Similarly, from Ireland, Roman Catholics excluded from Trinity College Dublin might go to Continental universities. But only in really active medical schools would they meet modern science with its body of more-or-less testable theory.

Priestley attended and taught at Dissenting Academies, which prepared him for his career in Unitarian ministry (and chemistry, which he taught himself), in the course of which, after a period of aristocratic patronage, he became a prominent member of the Lunar Society of Birmingham.[4] This group of friends included prominent industrialists such as Josiah Wedgwood, James Watt and Matthew Boulton. Wedgwood was a Fellow of the Royal Society and had invented a pyrometer to measure the high temperatures in his kilns. His attractive, consistent ware attracted much attention, and became fashionable. A standard-sized ball of standard clay was placed in the kiln for a standard time, and dropped down a grooved, narrowing and calibrated metal scale. It would have contracted; and the hotter the kiln was, the further down the ball fell. It was impossible to connect this scale with ordinary thermometer temperatures, because glass would have melted; but for potters and metallurgists it provided a form of process control. Whether it was science is an open question: not probably in the sense in which we would use the word, but Wedgwood was certainly seen as a philosophical manufacturer; and 'science' is not some Platonic entity changeless over time and place, but must be understood in context – and in the eighteenth century, that meant natural philosophy and natural history.

Watt, trained as an instrument-maker and working in the University of Glasgow, learned about steam engines because he had to repair the small-scale Newcomen engine used in demonstrations there. Perhaps picking up 'cutting edge' ideas about latent heat from Joseph Black, teaching chemistry there, he noted that a great deal of ironmongery got very hot and rather cold on each stroke, and devised a much more efficient engine in which the cylinder would always be hot and a separate condenser always cold. For this to work, pistons needed to fit much better in cylinders than was normally then achievable; and it was Boulton in Birmingham, doing precision castings, who made these engines practicable, and marketed them particularly in Cornwall where tin mines needed draining, but the coal had to be brought from Wales and was expensive. Again, although Watt and Black have been called partners in science,[5] the place of our kind of science in this story is again doubtful. To spot that existing engines were wasteful and inefficient, to devise a better version, and to get it made, was a great practical achievement. But as nineteenth-century physicists noted, the steam engine contributed much more to theoretical science than science to the steam engine, for ultimately its analysis led to thermodynamics.

After his house in Birmingham was sacked in the 'Church and King' riot of Bastille Day 1791, Priestley went to London to teach at the Dissenting Academy in Hackney. He hoped for support from the Royal Society, but was not warmly welcomed. He was a notorious firebrand and radical, in a time of political reaction in response to French war and terror. Banks, presiding over the Royal Society, steered a careful course and (excluding party politics and denominational religion) kept science flourishing and important through the years of war. But its importance was chiefly cultural: what went on in Banks' learned empire had little effect upon the war's outcome, and was not generally perceived as bearing upon the war effort. Steam engines, in contrast, fired the British economy so that allies could be funded, and the French defeated – as Sadi Carnot, son of Napoleon's chief of staff and pioneer of thermodynamics, recognised.[6]

Banks had made his reputation in sailing on Captain Cook's first voyage around the world, on HMS *Endeavour*; indeed it was he who had botanised at Botany Bay. The voyage gave him a strong feeling for global science, and he became a great promoter of empire.[7] His vision was Baconian, and the science he promoted was useful knowledge. We get a good idea of it from the splendid letter he wrote to Davy after the invention of the safety lamp for coal-miners when mines, getting deeper still, ran into explosive natural gas – the Royal Society as such had played no part in the research:[8]

> Many thanks for your kind letter, which has given me unspeakable pleasure. Much as, by the more brilliant discoveries you have made, the reputation of the Royal Society has been exalted in the opinion of the scientific world, I am of the opinion that the solid and effective reputation of that body will be more advanced among our cotemporaries of all ranks by your present discovery, than it has been by all the rest. To have come forward when called upon, because no one else could discover means of defending society from a tremendous scourge of humanity, and to have, by the application of enlightened philosophy, found the means of providing a certain precautionary measure effectual to guard mankind for the future against this alarming and increasing evil, cannot fail to recommend the discoverer to much public gratitude, and to place the Royal Society in a more popular point of view than all the abstruse discoveries beyond the understanding of unlearned people.

It is precisely from these 'abstruse discoveries' that we today expect new remedies and other advances in technology, because we have come to expect a different rhetoric.

The call for 'applied science' had come from Davy, Priestley's chemical heir, and the protégé of the Watts and Wedgwoods. Abandoning his left-wing sympathies on being appointed to the Royal Institution, he had in his inaugural lecture of 1802 hailed the coming bright day when chemistry would be harnessed to improvement.[9] His research in his first years at the Royal Institution, at the behest of the management in which Banks was prominent, was on tanning and agriculture. Here he did what other men of science had done, when investigating gunpowder or natural products such as opium or Jesuit's bark – he sought to find a scientific rationale for what empirical study had demonstrated to be good practice. His sensible recommendations about tanning (do not use too strong solutions, or seek to speed up the process too drastically; be prepared to use substitutes for oak bark, such as the Indian plant catechu) vindicated what went on in the tannery of his friend, and Coleridge's, Thomas Poole of Nether Stowey. Davy was awarded the Copley Medal of the Royal Society for this work, which was not superseded for half a century, and provided a chemical basis for a very important industrial art, leather being a very important commodity in the days before rubber and plastics.

Rather than being applied, the science came second to the craft or technology, providing a rationale and the basis for improvement, but was nothing to boast about greatly. Davy also delivered lectures on agriculture, subsequently published in a handsome quarto (1813), and then in a cheaper octavo second edition.[10] Here again science was invoked, both geology and chemistry, in what became a standard work for a generation; but, once again, it supported good practice with a scientific rationale, for example over manuring with fresh dung. The book sold well, and was still selling when Davy's *Collected Works* were published ten years after his death; so it was divided between two volumes of that set, in order not to compete.[11] Davy also participated in experiments on grasses for feeding cattle, done on the Duke of Bedford's estate at Woburn – where his health was drunk at one of the celebrated sheep-shearings, annual festivals of agricultural improvement. Here was a promise of plenty.

The Board of Agriculture, under whose auspices Davy gave these lectures, had been set up by William Pitt to improve food production through information and emulation. The 1780s and 1990s were years of population explosion (Malthus published his famous *Essay* in 1798[12]) and of poor harvests – which had been one of the triggers for the French Revolution of 1789 ('Let them eat cake'), but also caused great hardship in Britain. Poor rates were high, and hunger widespread: young Faraday's family may have been among those who needed handouts,[13] and the Quaker pharmacist William Allen was prominent in organising relief for the poor.[14] Reports on the agriculture of the counties of Great Britain were prepared in a series organised by the energetic booster Arthur Young,[15] Secretary to the Board. Young was an enthusiast for agricultural improvement, and a shrewd observer of rural life as well as of agricultural practice, although his own farming had not been successful and he was not personally responsible for any great innovations. Local gentry wrote the various volumes, which are thus even more various than the farms they describe; but what comes through is some of Young's enthusiasm, the realisation that the best practice will give much higher yields, and that this is genuine local and empirical experience rather than theoretical knowledge. That was what appealed to the conservative landowners and farmers to whom the series was addressed.

Through the years of war and blockade, unpromising land was brought into cultivation, and improved crops and livestock were more widely used. William Marshall, an equally energetic and forceful agriculturist, distilled these county volumes into five stout tomes[16] which preserve some of the down-to-earth tone of the best originals; they also indicate that, while farming in these years did absorb some of the ethos of industry, and its apologists foresaw plenty and profits, again it was empiricism, learning from one's own and others' trial and error and being open-minded, which was responsible, rather than scientific training. Davy's mildly more theoretical approach was often looked at askance; science may have been a gift horse, but shrewd farmers looked in its mouth. Ballyhoo could be resisted: the case for science as the basis for techniques was still not made.

There were, from the eighteenth century, local agricultural societies: clearly, in agriculture what is possible depends upon local conditions, and metropolitan societies or Boards are inevitably at a disadvantage in a way that does not happen in chemistry. We associate eighteenth-century Bath with frivolity and fashion, and with illnesses for which the water was the last hope of a cure – indeed, the Abbey is filled with memorials to those who came to Bath to die. But it was also the centre of a flourishing Society for the Encouragement of Agriculture, Arts, Manufactures and Commerce, founded by a group of gentlemen from the west of England in 1777:[17]

> The scheme received immediate approbation and great encouragement, not only by liberal subscriptions, but also by many useful communications of knowledge, both scientific and practical, from ingenious and sensible correspondents.

The scientific and ingenious are nicely contrasted to the practical and sensible: this is a very English tone. Unlike many less well-placed societies, these prosperous gentlemen were able to publish their ideas, in a journal which was so successful that the volumes went through a number of editions. In selecting and editing what had come in:

> Regard has been principally had to such as relate to matters of practice. Useful hints, however, of the speculative kind, which may in their consequences lead to practical improvements, have not been neglected; – such will always be esteemed valuable communications, altho' inferior to those that have already been submitted to the test of experiment. In a work of this kind, to be explicit and intelligible, are all the requisites with respect to language; and therefore the thoughts of our correspondents are generally given in their own words. The Society, however, think it necessary here to observe, that although they have no cause to distrust the knowledge or veracity of any person who has favoured them with his correspondence, yet, for obvious reasons, they do not mean as a body to vouch for the truth of any relation, or to give authority to any opinion contained in the following papers, further than the *notes* express, and to recommend them as subjects of enquiry and examination.

This was the spirit of Banks and his gentlemanly Royal Society. In the volumes that the society published, however, we can see optimism almost amounting to enthusiasm for science. There are letters about 'setting' wheat in Norfolk and Suffolk, where the best crops came from; on the great increase in milk from feeding cows sainfoin; on the cultivation of rhubarb, a plant important in medicine as a laxative, hitherto imported from the East; on cultivating trees on moors; on pest-control; and on machinery for draining land. The second volume quoted 'an ingenious writer':[18]

> Improvements in Tillage arise, in general, from the slow operation of doubting experience among men who obtain their bread by the sweat of their

brows, whose minds are not sufficiently enlarged to adopt, but with reluctance any deviation from the practice of their forefathers, and who are fearful of risquing the moderate certainty they possess for the prospect of greater gains which are unknown.

It was therefore appropriate that gentlemen, especially those of considerable landed property, should offer premiums for improvements insuring small farmers against loss through experiment, and should themselves experiment also. They thus bring lustre upon themselves, and are entitled to the nation's gratitude:

At a very trifling expense they become the primary means of increasing the wealth and happiness of the community, who feel, through every rank and order, the beneficial effects of every improvement that tends to increase the value and produce of our lands.

The ministers of the established Church of England, also, educated men scattered through villages, towns and cathedral cities, and whose income was largely based upon rents and tithes, should be involved in their own interest as well as that of their parishioners:

To the Beneficed Clergy also, of every rank, the encouragement of such Societies [as this one] ought to become the object of speedy and general attention. They are essentially interested in whatever tends to promote the improvement and value of lands, being sure to partake of the increased produce, without the lest loss from the failure of any experiment. It therefore seems incumbent on them to lend their assistance in supporting an institution which must increase the value of their livings in proportion as its exertions become beneficial to the publick.

This conservative rhetoric of improvement through empirical science, Banks' strategy for the prospering of science in revolutionary times, was just what Davy was to pick up in that famous inaugural address: while bringing prosperity to all, it need not threaten the rank and order upon which social stability depended. He has been seen as the apostle of applied science.[19]

Davy undertook soil and mineral analyses for a fee, and landowners could thus know what crops were likely to flourish, what fertilisers would be advisable, or what mines might be profitable. Work in the laboratory, depending upon inductive inferences, thus guided what was done in the fields. But his safety lamp research was presented as unlike what had been done before.[20] He had samples of the explosive gas from the pits sent to him in London, and in the laboratory of the Royal Institution he analysed them and found that they were methane. Investigating mixtures of this gas with air, he found that explosions would not pass through narrow tubes; and described to Banks, and to the committee in Newcastle, a lamp with such tubes at the top and bottom – which went out when the concentration of methane was dangerously high. Later, he discovered that wire gauze would have a similar effect, dissipating heat rapidly, and the classic design he came up with at the end of the experiments was a cylinder of such gauze in place of the glass in an ordinary lamp, surrounding the flame. When methane was present, the flame got bigger but did not pass through the gauze and the pit could be evacuated. The lamp was thus both a detector of gas, and a light to work by. The scientific principle was that heat was rapidly dissipated by the metal gauze, so that the ignition temperature of the methane and air mixture was not reached. Davy went on, using safety lamps, to investigate what came to be called heterogeneous catalysis, where in the presence of a metal (often platinum) a reaction will go smoothly at a low temperature rather than explosively. Wire gauze was also soon used in oxy-hydrogen blowpipes, producing a very hot flame, where it prevented the flame blowing back and igniting the hydrogen reservoir. For his work, Davy received a set of gold and silver plate from coalowners, letters from colliers, ennoblement as a baronet, and recognition from Flanders, Russia and other countries with mines. He thought that the usefulness of high-powered metropolitan science was proved.

But the story was more complicated than this.[21] Explosions did not become a thing of the past, though they did not rise as fast as coal production. Some were due to carelessness by miners or managers; but coal dust turned out to be also explosive, and lamps needed to be modified so that the flame was enclosed in glass and not just gauze. Also, others less eminent than Davy in the scientific world had devised safety lamps. William Clanny, a GP in Sunderland, had sent to the Royal Society, through the benevolent William Allen, a description of his lamp, in which the air entering and leaving the lamp was pumped through cisterns of water to cool and contain it.[22] This was a clumsy device, bulky and needing a boy to pump it continuously, and hardly seems to have been adopted in practice. Clanny's chemistry was a little shaky: the explosive gas is sometimes (correctly) 'carburetted hydrogen', sometimes 'hydrogen' and sometimes 'inflammable air'.

Meanwhile, at exactly the same time as Davy, George Stephenson, who was to become famous for his locomotive engines, but in 1815 was still a colliery foreman on the Tyne, unknown beyond his small circle, also invented a lamp, later called the 'Geordie'. He sought to make lamps safe by restricting the access and egress of air, using narrow tubes or tinplate with holes punched in it; and his lamps retained the glass around the flame that ordinary lamps had. He and Davy were working simultaneously: the practical man and the eminent chemist, both improving their designs that autumn essentially by trial and error (though Davy knew for sure what the gas was, and had a theory of heat) and aware of the other's work. Davy, wealthy from his marriage and disposed to scientific philanthropy in the tradition of Rumford and Allen, refused to patent his lamp, so the question of priority was never tested in the courts. To Davy and his supporters in London, it was evident that a man ignorant of science could not have made the invention, and he intemperately accused Stephenson of piracy; but Stephenson had many supporters in the North, who regarded him as hard done by, and Davy as muscling in and claiming credit for another's work.[23]

Davy, like Newton in his dispute over the Calculus with Leibniz, got Banks to set up a committee of the Royal Society to vindicate his claim, and do down Stephenson. When in 1820 Banks died, Davy as the gentleman with the lamp was the obvious candidate (despite his lowly provincial birth) to become President – though some of the Fellows resented being directed by such a person. Things got worse when the Navy wanted advice about preserving the copper bottoms of their wooden warships. The copper kept out the worms but was slowly corroded. Davy saw how to deal with this problem: in his work on electrochemistry, he had demonstrated how a metal in close contact with another which was more reactive would acquire a negative charge and become inert. This which we call 'cathodic protection' involved the latest chemistry, for which Davy had received a prize from the French Academy of Sciences. In a tank in the laboratory, copper in contact with iron remained bright when immersed for a long time in salt water, while copper on its own was corroded. When the experiment was scaled up, to a ship in harbour, it still worked – but unfortunately when iron 'protectors' were ordered for the entire navy, it turned out that at sea marine organisms, which had been repelled by the slowly dissolving copper, now adhered so strongly to the ships' bottoms that they sailed badly.[24] Davy was greatly mortified, becoming the butt of cartoons and stories; and the bad temper he had shown in responding to Stephenson became more evident. He had to retire early from science and from the Royal Society's presidency following a stroke, and he died abroad at the age of fifty. On his death-bed, he had been experimenting with a naturally-electrical torpedo fish, to see if he could induce electrochemical effects with it. The practical application of the latest science, notably the scaling up of experiments, was and remains a difficult business: Davy, the Mr Fixit of his generation, could not in the end quite bring 'applied science' unambiguously into the public eye.

Nevertheless, the nineteenth century did become the age of science,[25] because a way was found of linking metropolitan and provincial skills and resources. Because Germany was not until 1870 a nation-state but a mass of large and small principalities, Germans lacked the critical mass necessary to propel modern science in the 1810s onwards. Vienna, Jena,[26] Weimar, Munich and Berlin at different times were important centres as people came together there – but then these groups broke up. Lorenz Oken had the idea of calling meetings of all the Naturforscher of Germany together in a different state each year: at first the authorities looked askance at such congregations of intellectuals who might think dangerously, but soon realised that they were fairly harmless and that hosting a meeting could bring prestige. Foreigners from outside the German states were also welcomed. German was not much spoken in Britain, but among these guests were two Scots, David Brewster and James Johnston, who were both delighted by what they saw and heard.

Meanwhile, Davy, Wollaston, and the polymath Thomas Young, the old guard of British science, had all died in 1828–9. Although we know that Faraday was embarking on the electro-magnetic researches that would transform everybody's life by the twentieth century, and Darwin was setting sail on HMS

Beagle, it seemed plausible to see (with Charles Babbage) British science in decline.[27] In 1830, Davies Gilbert, Davy's patron and then successor as President of the Royal Society, resigned. Two candidates emerged: John Herschel, who was backed by the cantankerous Babbage, and the Duke of Sussex, the most intellectual son of George III. Babbage canvassed hard, but was over-confident and thought the result was in the bag; and, in the event, the Duke won. The majority of Fellows evidently wanted a President like Banks, only more so: a genial nobleman with access to all the corridors of power, allying science firmly with rank and fashion, rather than a man distinguished for outstanding research.

Most of the active minority of Fellows had wanted Herschel. They hoped, as indeed Davy had, for something more like an Academy of Sciences than a club for intellectual gentlemen: Davy had been amongst the founders of the Athenaeum in London, making Faraday its first Secretary, to fulfil the latter purpose. Although in the event this is what happened, and the Duke's Presidency was not reactionary, there was deep dissatisfaction in the Royal Society at his election. London science seemed to be vitiated by snobbery and what we would call amateurism. But an exclusive Royal Society would have serious disadvantages, notably in communicating and disseminating science, and in diffusing an enthusiasm for it: and while London was a good centre, wide access, and the expertise and assistance of those based in other places, were needed too in forwarding science and its reputation for practical usefulness.

In York, the Reverend Vernon Harcourt, son of the Archbishop of York, was interested in science (having set up a laboratory with advice from Davy and Wollaston) and presided over the Yorkshire Philosophical Society, which held notable fossils from Whitby and Kirkdale, where Buckland had identified a hyena's den perhaps destroyed in Noah's Flood. York, a great cathedral city and administrative centre of a rich region, was becoming, through this society and its major figure, the lecturer and curator John Phillips, an important provincial focus for science;[28] and was to become a hub of the railway system. Harcourt saw that an institution like the German one, meeting annually in a different provincial city, might be just what was required, and he convened one in York in 1831. A bereaved Buckland couldn't come at the last moment, but enough people were there to make it a success; and they were invited to Oxford the next year, when Buckland ensured that doctorates were conferred upon Faraday, Dalton and others. In York, they hit upon the idea (now familiar from big conferences) of parallel sessions running simultaneously but devoted to different sciences. These 'sections' became a feature of meetings, and a way in which participants identified themselves.

Meetings were open, for all men and women who could afford to pay and to take time off. Local worthies were appointed as Vice-Presidents, and while 'gentlemen of science' from London, Cambridge and Oxford ran the show, locals could and did also present papers there in the appropriate sections. Well-reported meetings in industrial and commercial cities followed those in university towns, and became a feature in the calendars of those interested in science;

and lectures were given to 'working men' by some of the scientific lions who formed the platform party. Cities competed to hold meetings, promising as part of their bid to build a museum, public library or mechanics' institute, and hosting grand receptions. Sometimes plenary sessions were actually held in theatres; there was always something theatrical about the occasion. The President, a distinguished man of science (ideally with some local connection) held forth about the triumphs of science in the last year, and its promises for the future, provided that more support from government was provided. Usually, a volume about the city and region was published, drawing attention to its history, resources and potential. Behind it all lay the promise that knowledge was power, and that applied science would transform society.

Sometimes there was some legislating to be done: Berzelius' chemical symbols with which we are familiar (O for oxygen, H for hydrogen and so on) were approved over Dalton's hieroglyphics; proper rules of priority in agreeing binomial Latin names for plants and animals were agreed upon; and heresies like the 'quinarian' system of circles in taxonomy were vigorously denounced. All this was news. Sometimes (to the delight of some at least of the visitors) there were great rows. The best-known example is the confrontation between Bishop Samuel Wilberforce and Huxley in Oxford in 1860, an event which has since acquired an epic significance that it did not have at the time.[29] Huxley was involved in other battles at that meeting and others, giving some of the excitement of a prizefight to what might otherwise have been tediously over-decorous proceedings. Distinguished foreigners might be there; laudatory speeches (*treacly*, mutual be-buttering, abominable speechifying and flummery[30]) were the order of the day; and a good time was had by all. Indeed, then and still nowadays, the meetings were an excellent opportunity to take an intellectually stimulating holiday, being entertained by guest lecturers, meeting old friends and seeing local sights on guided tours. Such breaks were popular with the serious middle classes, accustomed to taking long holidays; as were times at the seaside, arranged notably by Philip Gosse, where marine life could be studied in rock pools, and an aquarium assembled to take home.[31]

Abroad, the BAAS was the prototype for similar groups in the USA, in Australasia and in France. At home, other bodies, such as the Royal Agricultural Society, followed the same plan of meeting in different towns. In 1844, for example, in the 'hungry forties', it met for the last week of July in Southampton, and published for those attending a handbook[32] with the programme, and suggestions for those who might spend the following week there on 'tours of pleasure and observation'. The map prominently indicates the railway station: the coming of railways made these large peripatetic meetings possible, greatly reducing the tyranny of distance. On the Wednesday, there would, at the Victoria Rooms, be 'a hot dinner ... a magnificent thing; it is supplied by Breech and Co. of the London Tavern, and will comprise every delicacy that nature assisted by art can produce.' There the prizes would be awarded; and the doors would open at four o'clock. On Thursday, an hour earlier, there would be a cold banquet in the Pavilion, large enough for 1,200 people, with a gallery for ladies. Other attractions, apart from the exhibitions of livestock, crops and implements for which most must have come, included the new hospital, a new mercer's shop in the Saracenic or Moorish style, a picture gallery, and a theatre (no mere music hall) where 'legitimate drama, opera and ballet are placed on the stage with a liberality and correctness of taste, unapproached out of the metropolis', Reading Rooms, where a painting of the previous meeting (at Derby) was displayed, the Masonic Hall, the public baths, the docks, and trips by train to Winchester and by steamer to the Isle of Wight.

A scientific enterprise in progress but clearly needing a little public support was the digging of an artesian well, which had not so far answered the expectations of its backers, although it was now 1,250 feet deep:

> As all objects of any great enterprise have individually been subject to public distrust in their progress, and have but seldom yielded pecuniary profit to their promoters, it behoves all those who regard our national fame for science to inspirit rather than discourage the boldness of those who speculate in such undertakings as the Southampton Artesian Well.

The Society's support would be worth having, for it boasted an even more socially glittering list of Vice-Presidents than the BAAS could do, including three Dukes, a Marquis, four Earls, two Lords (one a Member of Parliament) and a baronet. Its Council further included seventeen MPs. It also retained the services of a consulting engineer, Josiah Parkes, and a consulting chemist, Lyon Playfair.

London was not completely forsaken by those in search of science, however, and Playfair was one of the crucial figures behind the Great Exhibition of 1851, under the auspices of the Society of Arts but strongly promoted by Prince Albert.[33] This was to illustrate the works of all nations: there had been a number of regional and national exhibitions, but this was to be the biggest and best. Official guides, and the heavyweight reports of the international 'juries' set up to compare the various raw materials and manufactures exhibited and award prizes, were published; but there was a plethora of other publications that marked this extraordinary event. Lyon Playfair and Henry Cole were keen that lessons should be drawn from what had been on show, and were uneasy about the general smug satisfaction that British had been shown to be best. Cole noted the elegant industrial design apparent in France; Playfair the ingenuity and mass-production of the American exhibits, especially the Colt revolvers. In the aftermath of the Exhibition, and associated with it, was a series of public lectures[34] at the Society of Arts, delivered weekly between 26 November 1851 and 3 March 1852, in which eminent men of science drew conclusions from what they had seen there.

William Whewell drew attention to the way science generally followed arts, just as the critics and grammarians came later than the poets of Antiquity. This, despite the example of Davy's lamp; and the lectures abounded in discussions of the relations between science, involving theory, and practice. These were at that time often set in opposition:[35] but these lecturers did their best to harmonise them. Whewell (*Lectures*, p. 33) used his word 'scientist' (with an apology) in contrast to 'artist', someone making something useful and/or beautiful. Other lecturers were Henry de la Beche, Director of the Geological Survey; Richard Owen, the palaeontologist; Jacob Bell, an MP and pharmacist; Playfair; John Lindley, the botanist; Edward Solly, FRS; Robert Willis, mathematics professor and expert on tools; James Glaisher, meteorologist; Henry Hensman, civil engineer; J.F. Royle, whose expertise was on the natural history and natural products of India; and Captain John Washington, FRS, and soon to be appointed Hydrographer to the Navy and an Admiral. Lindley and Solly took plants used as food, and in industry, respectively.

Although the Exhibition was the occasion for a great deal of ballyhoo, and of genuine enthusiasm, patriotic pride, and insights into the power and ingenuity of mechanical inventions, making belief in progress widespread and reasonable, the lecturers were on the whole surprisingly sober. They drew attention to instructive failures; to the poor state of British education, demonstrated in ignorance and conservatism; to the essential importing of experts from better-educated Scotland and Germany; to the way they did things better and more elegantly in France; and to the lack of state support for the science upon which future progress, and even the British economy holding its own rather than declining, would depend. Washington noted that in Napoleonic times, captured French ships were superior to those built in Britain, where traditional practices and copying prevailed. Clearly, those who bemoan the decline of British industry are in a long historical tradition.[36] Washington referred to the hopes of peace and prosperity that the Exhibition was intended to foster, and to its success; but also drew attention to the armaments exhibited there. Warfare at that time, including the winning of the American West, was (as since) a great promoter of technical development, source of anxiety and pride, and a sphere in which money was no object.

The Exhibition had cheap and expensive days, effectively thus including artisans as well as the middle classes in its public. Audiences at the lectures would have been not unlike those at the Royal Institution, or the fully paid-up attenders at the BAAS: we are here encountering the thoughtful and serious. Others must have seen the occasion as a huge success, marvelled at millions of visitors, the gallons of tea and hundredweights of refreshments sold there and the profit made, and rejoiced that the Crystal Palace was re-erected (slightly modified, and surrounded by huge models of dinosaurs) in leafy Norwood, where it remained until burnt down at the end of 1936. We are amazed at the rapidity and efficiency with which the whole Exhibition was planned, in months rather than years, and was ready on time (except for the American exhibits). Railways made the construction and the huge numbers of visitors possible. Contemporary commentators were staggered not only by what was on view, but also by the way that the imagination of the orderly crowds of working-class families was caught. A harmonious world, rather than a struggle for existence, seemed to be on the cards, as Davy had said half-a-century earlier. Technology, especially in its new alliance with science, seemed essentially benign. Although 1848 had been the year of revolutions across Europe, in Britain, where the large Chartist demonstration in Hyde Park in 1848 had been peaceful, and the Exhibition so successful, fears of revolution at last abated.

Subsequent exhibitions around the world never quite equalled this first one, either in enthusiasm generated or profitability: the Commissioners used their funds to buy an estate in Kensington to be developed as a cultural centre, and there was still some money for scholarships. Later exhibitions were organised on this site: in 1862, clouded by the death of Prince Albert, and housed in a very non-magical building; and in 1876, when a display of scientific apparatus was accompanied by another series of lectures, to which we shall come later. By then, the Prussian victory over France in 1870–1 had given an enormous stimulus to scientific and technical training. The world was rapidly specialising, and these lectures were published in three volumes:[37] two devoted to the day-time sessions for the middle classes, and one to the free lectures given in the evening to a more mixed audience. Focused upon historic equipment, used for example by Dalton and Faraday, and also upon the latest apparatus, the sessions reveal the drive for accuracy and quantitative measurement so important in this era, the reign of classical physics; and aimed to generate informed enthusiasm. Playfair was again a major figure.

By then, Cole's vision of a museum of design, and a college of art, were being realised: the site as it was developed over the next twenty-five years was also to include the Albert Hall, a music school, the 'normal school' for training teachers, and the Imperial Institute to display colonial products; while, adjacent to it, was the Royal Geographical Society. The design museum which hosted the sessions became the Victoria and Albert, where an amazing ceramic staircase commemorates Cole, and displays his vision of science and art in harmony; the Natural History Museum (where again the very architecture indicates the contents and their sublime importance) was separated off from the British Museum in Bloomsbury; the Science Museum, where some of the apparatus from 1876 ended up, was built; and the Normal School, School of Mines, and Royal College of Chemistry were amalgamated into the Imperial College of Science and Technology. Technical education in Britain was also greatly stepped up in the years around 1900: the idea that artisans (like doctors) needed formal training as well as on-the-job experience had prevailed, and there were (as there had not been in 1851) major science-based industries concerned with synthetic dyes, and with electricity. There was no doubt any more whether science was an important part of culture, the educational system and the economy.

In Edinburgh, the university had in 1855 put its museum in the charge of George Wilson, a chemist who had worked with Playfair, appointed to the new post of Regius Professor of Technology. Although the word goes back to the seventeenth century, he had to spend much of his time explaining what his subject was supposed to be – and essentially urging the importance of science in industry.[38] Wilson was an invalid, dying young in 1859; and the university failed to appoint a successor. But for students there, the choice of a scientific career as new professions opened up was becoming possible. For the working men Huxley addressed at the School of Mines, who came to (or read afterwards) lectures at BAAS meetings, at the 1876 sessions, or at Mechanics' Institutes in mid-century, there was no such hope: science for them had to be an interest, an opening of windows. To be encouraged to think empirically, to follow what was described as scientific method, must have been useful educationally; but it was only late in the century that technicians with scientific training were required in any numbers. The word 'technician' unfortunately already carried overtones of narrowness, as specialisation became the order of the day and the gulf between scientific writing for scientists and for outsiders widened.

We shall come to display next. In exhibitions and museums, the hope is that the public with the fruits of the earth and of science before their eyes will come to the right conclusions. But naturally they often do not: here as elsewhere, people are awkward and like to come to their own conclusions rather than someone else's. Curators' attempts to steer those who come to look are interesting but not always effective; people are counter-suggestive. The same happened and happens with reading: we have always read into texts what we would like to see there, and memory is also a creative faculty. This is not something new that came in with post-modernism: Isaiah's 'sign' of a young woman soon to bear a child and call him Immanuel became, for the first Christians, a prophecy of the virgin birth of Jesus,[39] in what we could call a creative misunderstanding (as modern translations make clear). It remains a problem that we know more about writers than readers, except for the small and unrepresentative class of reviewers, though all writers have in mind an ideal reader; and the same is true in the realm of spectacle.

Books, periodicals, museums and lectures have it in common that they are a part of scientific practice as well as attempts to get knowledge across: what scientists did and do was to write, exhibit and speak as well as experiment and theorise. In studying religion we get a very curious picture if we desert practice, what people do, in favour of what they supposedly believe: that way, one cannot get inside those involved. So with science, as we turn to display, we should remember that this was in part the practice of science, and only partly a way of getting the general public enthused. In museums, botanic gardens and zoos there are and were in the past unavoidable tensions because different people make different demands and have different needs. This is not simply a gulf between the professional and the popular: just as Mary Somerville wrote books that helped specialists understand what others in different branches of science were up to, so in display there is no great gulf fixed between communication to the educated and the lay. For everyone display (like ballyhoo) was and is more fun than learning a syllabus.

7 Display

Robespierre's amazing Festival of the Supreme Being proved with panache that there was a new order of things to go with the new calendar, with its revised months and ten-day weeks. The French revolutionaries after 1789 needed festivals and displays to maintain morale in the face of inflation and military threats, and laid on patriotic alternatives to the great public occasions of the ancien regime. When Napoleon seized power, he also proved a master of ceremony and display, the high point being his coronation, where he took the crown from the captive Pope and put it on his own head. Triumphal arches, buildings adorned with his monogram, splendid uniforms for the Imperial Guard, and ermine for himself brought legitimacy to this upstart from Corsica, who like the hero of a fairy tale had conquered continental Europe, and married a princess.[1] Science was still in 1789 itself an upstart, better-established in France with its paid Academicians than elsewhere, but marginal in the old world ruled by churches and kings, where it was essentially an interest for gentlemen with as yet little practical outcome.

The revolutionary government under Robespierre and the Jacobins during their reign of terror abolished the Academy as an elitist and reactionary body, and executed Lavoisier, effectively its head, in 1794. But the expertise of chemists and engineers, working like Lavoisier on the improvement of gunpowder, on surveys, and also on making guns, proved essential to the state, defending itself from foreign armies, and then exporting revolution, bringing caps and trees of liberty, and guillotines, to those liberated willy-nilly from oppressors. So the Academy had to be in effect re-founded, Napoleon being elected, as the First Class of the Institute.[2] What had not been suppressed was the Jardin des Plantes on the left bank of the Seine, the great botanic garden where taxonomists had improved upon Linnaeus' method, using a natural system based upon many characteristics rather than his artificial sexual system. This entailed a museum collection of dried specimens, for study and comparison. The animals from the King's menagerie at Versailles were brought to Paris, where the zoo, open to the public, still adjoins the botanic garden, and where some buildings go back to the revolutionary epoch.

The revolutionary closing of elitist organisations, and suppression of licensing and professional bodies which had excluded the unqualified, threw open medical careers for those who before could only have aspired to low-status jobs in the field of health.[3] At the Jardin des Plantes and its associated Museum, professors gave free public lectures aimed, unlike most of those at the Royal Institution, to prepare medical students as well as interest the general public. At the revolutionary (and increasingly militarised) École Polytechnique, the carefully selected students were taught by active researchers in mathematics, chemistry and natural philosophy in a great step towards the modern elite university; but, at the Museum, the lectures by researchers no less distinguished for their research were open. When peace finally came after the Battle of Waterloo in 1815 and the subsequent Congress of Vienna, Britons flocked to Paris in the wake of the Scottish author John Scott,[4] and keen students of science and medicine sought to catch up with what had been going on during the war, and hear the great men who had kept France at the forefront. In zoology, the most eminent were J.B. Lamarck and Georges Cuvier; and Cuvier went on to become the permanent secretary of the First Class of the Institut, flourishing and increasing in power and prestige under the Napoleonic and then the restored Bourbon monarchy.

Scott reported on the important place that the sciences occupied in Parisian culture, on the boost that Napoleon (like later tyrants) had given to scientific and technical training at the expense of education in humanities. His informant noted the chemical laboratories attached to scientific institutions, and the many scientific professorships – though noting ominously (*Visit*, p. 233) that, while since the Revolution,

> They have been undoubtedly improved by new organizations, and by the impulse which has been given to the physical sciences in general. The salaries are all paid by the government, and they are very moderate.

This last point was to lead to the practice of 'cumul', in which leading practitioners held a number of posts, assigning the duties to ill-paid juniors, like eighteenth-century rectors and their curates in English parishes. But in 1814 all seemed well, and the museums of Paris, where knowledge was indeed on display, were particularly noteworthy to tourists. The Louvre still contained all the loot of Italy, Germany and the Netherlands brought home by the French armies,[5] and only after Napoleon's hundred days and Waterloo were the works of art returned to the homelands of the victorious allies, along with some, but not most, scientific collections, in a less-indulgent peace. Scott also included a description of the Jardin des Plantes, with its beautiful botanic garden, its restaurant, its zoo, and its museum, the richest in specimens in the world, though its library was perhaps inferior to Banks'. All was well-lit and well ordered; the zoological treasures were (*Visit*, p. 297):

> Contained in the long gallery on the second [floor]. They are well-lighted by semicircular windows in the roof. The length of this gallery, and the diversified and numerous assemblage of beings which are crowded in it, form a

pleasing and animated coup d'oeil, and the interest heightens, when, on public days, we find it nearly impossible to move through the crowd of persons of all ranks which fills it.

Such democratic display of knowledge was not yet a feature of English institutions such as the British Museum, which was essentially for ladies and gentlemen. Scott was also much interested by the Conservatoire des Arts et Metiers, devoted to applied science. There was no such permanent exhibition of the latest techniques in London, because the Royal Institution's original plan had not worked out. It was open to the public on Sundays only, but he was able to frequent it. There were spinning jennies, machines made by the celebrated Vaucanson, models of chateaux and manufactories, the basket from a pioneer balloon, fire pumps, agricultural machinery, distillation apparatus, a Chinese pagoda, and a large clock with an organ attached surmounted by a glass celestial sphere and orrery. This was thus one of the first museums of science and technology nowadays so attractive to children of all ages. If it seems surprising that it should have been in France, rather than in rapidly industrialising Britain, we should remember that British manufacturers were (prudently) sensitive about industrial espionage, enjoyed rather weak patent protection, and did not want to broadcast their trade secrets. Display can be looked at askance.

In 1823 Joseph Deleuze (translator of Erasmus Darwin) published, in Paris but in English, an account of the museum and the gardens, including maps and engravings of buildings and animals:[6] now, again, 'Royal' under the restored monarchy, they were becoming a tourist attraction. In describing its history, Deleuze referred to plans that had been made to place the museum and gardens under a single director, rather than a collective leadership of equals: this proposal had been resisted, and finally abandoned. The result was a little Garden of Eden:

> How pleasing amid the agitation of a great city to behold an establishment, in which are united fifty families, living in peace, usefully occupied, contented with their lot, attached to the place of their abode and priding themselves in its prosperity, strangers to professional rivalry and political dissensions, and grateful at once to the government which supports and the administration which directs them.

We may smile, as Davy did at young Faraday's notion of the moral superiority of scientists; but cannot doubt that this was both a very high-powered scientific centre, and an accessible public institution. And, in fact, French scientists in the revolutionary and Napoleonic periods, and right through the nineteenth century, were often prominent in political life, and not simply grateful to whatever government was supporting them. Deleuze particularly recommended botany for women:

> It presents nothing to offend their delicacy; it furnishes their amusement in retirement, and lends interest to their walks; attaches them to the cultivation

> of their gardens; assists them to develope a habit of observation in their children; and affords an opportunity of gratifying their benevolence, by making the poorer inhabitants of the country acquainted with useful plants. The letters of Rousseau first excited a taste for this science in the ladies of France, which has increased with the facility of obtaining instruction. A considerable number repair to the garden at an early hour to attend the lectures, and a separate space has been reserved for them in the amphitheatre.

The total audience at the lectures, delivered three times a week in the summer months, was, we are told, five or six hundred. Ladies were also taught iconography, the art of drawing natural history illustrations, with the professors on hand to indicate what was of scientific importance.

As part of what can be called the second scientific revolution, the science of geology had come into being quite distinct from eighteenth-century Theories of the Earth: the age of 'systems' (including evolutionary speculations) was past, and now positive knowledge was required. An important aspect of this was the reconstruction of extinct animals, where Cuvier was the world's leading expert, and the quarries around Paris, especially at Montmartre, were at first the most important site. Napoleon ordered that during his works in Paris, fossils found in the quarries should be brought to Cuvier's attention; and, in the museum, the fossils were under the same roof as the skeletons of animals still around. By comparing specimens, Cuvier had been able to reconstruct extinct creatures, establishing that there had been a series of Parisian fauna; those looking at the displays in the museum, where samples were on show and more in drawers, could do the same:

> The system which M. Cuvier has introduced in comparative anatomy has enabled him to determine to what genus even an insulated bone belongs, although the animal should have no living analogue. When he established the genus anoplotherium, it was from the scattered bones of different individuals that he determined the general form and distinguishing characters; a short time after, the almost entire skeleton was discovered, which we see above the cases, and it was found perfectly conformable to the description which he had given of it.

Cuvier's principle of correlation required great skill and experience, but was not too recondite for the visitor to be unable to appreciate its results when confronted with the various dry bones that Cuvier had brought back to life.

Among these were ichthyosaurs, some given by Buckland, and pterodactyls; but the name dinosaur was not yet coined, and the great skeletons which were later so prominent in museum displays were not yet excavated. Extinct mammals and fish (which became Cuvier's speciality) were the most obvious features of the fossil collections. Visitors would have got from the display some feeling of the length of the Earth's history, and of the various (and perhaps distinct) epochs in which different kinds of creatures had flourished. Although models of extinct creatures do not yet seem to have featured, drawings of them were being made and exhibited. Seeing the collection would be very different from reading about fossils; and the work of Cuvier in particular was leading to palaeontology becoming the leading edge of geology, as the strata could now be assigned relative dates from the fossils embedded in them.[7] Similar fossils were laid down at the same epoch; and, from comparisons, a geological column of all the strata (never found together in any one place) was being constructed. Fossils were like coins or medals dug up, both interesting in themselves and keys to the past. What could be learned from Cuvier's great book on the animal kingdom was made visible in the museum's displays.

Davy was one of the few Britons to have visited Paris when it was still the enemy capital; and when he became President of the Royal Society in 1820, he began the process of making it a little more like the Academy. This ambition also involved making London more like scientific Paris. One of his achievements was, in alliance with Sir Stamford Raffles, historian of Java and founder of Singapore, to found the London Zoo.[8] Raffles had, in 1817, 'meditated the establishment of a Society on the principle of the *Jardin des Plantes* at Paris', and while Davy looked more to the usefulness to country gentlemen of naturalising exotic species, Raffles was more concerned with 'the scientific department' of zoology. In Java, he had supported the researches of the American doctor, Thomas Horsfield, which resulted in a splendidly illustrated book in which, among other things, the Malay tapir was brought to the attention of the world.[9] Raffles died in his mid-forties in 1826, when the Zoological Society had just been founded, on land made available in Regent's Park; shortly afterwards, Davy had a stroke, dying in 1829 at fifty. These early deaths deprived the zoo of its eminent leadership; but it flourished, as both a scientific centre and a spectacle. Raffles' first great collections had been lost in a shipwreck, but he had made more and these went to the zoo.

The prospectus of the Society is a curious document, beginning with the statement that:

> Zoology, which exhibits the nature and properties of animated beings, their analogies to each other, the wonderful delicacy of their structure, and the fitness of their organs to the peculiar purposes of their existence, must be regarded not only as an interesting and intellectual study, but also as a most important branch of Natural Theology, teaching by the design and wonderful results of organization, the wisdom and power of the Creator. In its relation to useful and immediate economical purposes it is no less important.

It was a matter of regret and reproach that there was no zoo in London, as there was in most European capitals. Its model was the Horticultural Society (later to have important gardens in South Kensington) so that emphasis was put upon domestication and acclimatisation: 'it is impossible not to hope for many new, brilliant and useful results ... by the application of the wealth, ingenuity, and varied resources of a civilized people.' In ancient Rome, wild animals had been displayed in the Colosseum, brought there to destroy and be destroyed as a spectacle; in Britain, they would be applied to useful purposes or scientific research, not vulgar admiration. The public would however be admitted to the zoo; in the event, their admission-tickets, rather than the subscriptions of the nobility and gentry, and men of science, kept the enterprise in the black.

Just as animals had been brought in to the Paris zoo from the royal menagerie at Versailles, so the denizens of the Tower menagerie (where animals presented to kings and queens had lived or languished over the centuries) were brought to Regent's Park as a nucleus of the new zoo. Illustrated guides were soon produced. The Victorian zoo became a huge public attraction, with rides on elephants making it a great family day out. Whereas earlier generations had had to be content with prints of Dürer's rhinoceros, now there was one on public view. Lions, tigers, serpents and hyenas could be observed, and their feeding and other habits noted: the symbolic, fairy-tale aspects of animals (the chivalrous lion, the indomitable rhinoceros at war with elephants, the ape that imitated mankind) gave way to a much more down-to-earth vision. Again, seeing actual animals was a much more powerful experience than seeing even the excellent pictures of them that were a feature of the books of this time. What was clear, and especially so as great apes were added to the collections, was we were not so unlike the animals as some might have liked to think.

When exotic animals died, their corpses were made available for dissection to members of the Zoological Society. Sometimes they were eaten: in the siege of Paris in 1870–1, the animals in the zoo were killed and eaten, but that was in the interests of survival rather than culinary or zoological science. Frank Buckland, son of the eminent geologist and a great populariser of natural history, was, like his father, notorious for the various dishes based upon the flesh of unlikely animals that he was wont to serve his guests.[10] He also kept unlikely pets, and once, when travelling by train, had been told that he must pay for his monkey because it was a dog, but not for his tortoise because it was an insect. But just as equipages drawn by zebras, or herds of buffaloes or llamas, did not in the event feature much in Victorian Britain, so stewed alligator (which is rather rubbery and faintly fishy to my taste), bear or snake did not find their way on to menus; nor did the nourishing insects (not tortoises) which people were also exhorted to consume.[11]

Dissection (which might precede consumption) was a more seriously scientific business. The star performer was Richard Owen, curator of the museum of comparative anatomy at the Royal College of Surgeons, and subsequently Superintendent of the natural history collections at the British Museum, and in effect first Director of the Natural History Museum.[12] His reputation made with the reconstruction of the moa, the giant extinct bird from New Zealand, he became the 'English Cuvier' and, like the original, he depended upon having at his disposal a vast collection for comparison.[13] Though he read and enjoyed Tennyson and William Morris,[14] his own writings were mostly very dry, though the *Palaeontology* recently reprinted is readable by non-experts, and was sometimes given as a school prize. He annotated his books copiously, and there is a huge archive at the Natural History Museum.[15] But it is in that museum, and the older ones where he had been in charge, that ordinary people would have been able to learn from the exhibits on display what the 'Dragons of the prime/that tare each other in their slime' had really been like, and how the families of animals were arranged.[16]

For Owen and most of his contemporaries, as earlier for Cuvier, 'family' and 'related', when applied to animals and plants, were metaphors: the Darwins, Erasmus and Charles, were people who took it too literally, and followed will-o-the-wisp evolutionary hypotheses. Owen's museums therefore demonstrated change over time, because series of extinct creatures found in the different strata were displayed there; but, with Cuvier, he doubted, or perhaps felt it prudent to doubt, whether such change could have been continuous. Different epochs of stability were separated by catastrophes, after which new species immigrated or were created. For the various groups, there were archetypes,[17] ideal forms, for example of a crustacean. Archetypes were realised in nature in the form of barnacles, shrimps, lobsters and crabs – all of which had their various species, at different times and places.[18] None of the actual species corresponded exactly to the archetype, a generalised form, which nevertheless was in a different and perhaps deeper sense, real. This idea led to what became the standard way of teaching zoology, and also of organising museum collections. When at last he got the great Natural History Museum built, a cathedral of science indeed, he was rising eighty. Just before retirement, he could display exhibits as he wanted, though mortified through the machinations of Huxley into having a statue of Charles Darwin displayed on the great staircase. The arrangement of animals might not look very different when done in an evolutionary way, but the archetype being diversely and incompletely realised, contrasts with the more materialistic notion of community of descent, and divergence from less-specialised ancestors.

At the Natural History Museum, where Owen's statue now occupies pride of place, and which from its full opening in 1881 (though the transfer was not complete until 1883), became one of the great attractions of London, Owen's successor was William Henry Flower. He was a disciple of Huxley and Darwin, and at once set about using the museum to display evolution. He was President of the BAAS in 1889, and devoted his address to the discussion of museums, later publishing it and other discussions of museums with biographical essays on Owen, Darwin and Huxley, and a brief history of the zoo.[19] Crucial in his conception of the modern museum of about 100 years ago was the separation of instruction and research. The public display must not be too cluttered with examples; there, arrangement is more important than content. He saw both functions of museums as extremely important, writing about museums aimed at boys (fostering and directing propensities to collect) and for schools, as well as emphasising the work of classifying and naming that went on behind the scenes in the great national museums.

Naturalists in the colonies were expected to send their collections 'home' to be properly handled in the metropolis, rather as primary products were shipped to Europe for manufacture into finished goods.[20] Museums were thus a focus of professional training as well as public information, and paid curators were essential, even for regional or school museums. By this time, on the Parisian model of a century before, science museums were expensive centres of expertise, comparable to research universities, but more accessible to the public (free indeed in London) and thus democratic, bringing national prestige and forming important international links as they exchanged specimens and supported expeditions and collectors.[21]

Given the wealth of Victorian Britain, and its far-flung Empire,[22] the collections at the Natural History Museum became particularly full. National museums all over Europe, and in the USA, played a comparable role; this was indeed their heyday, and a wonderful route for the inquisitive to enter science – and under the careful control of men like Owen and Flower, eminent natural historians who had their quirks and preferences, but were overseen by boards of distinguished trustees, were prominent members of their profession, and were clearly in charge of popularising establishment science. These museums were and are, unlike some others run by showmen, unconcerned with freaks, oddities or crankiness:[23] the public got what its betters believed was right for it, and museums provide a good example of the classic vision of popularisation, getting across facts and ideas from up-to-date science in readily-appreciated form. That is no doubt why they can seem (and can be) stuffy and dull to the easily bored and the rebellious.

In Paris, the Jardin des Plantes gave its name to the museum, and is near the Sorbonne; and the zoo adjoined it. In London, the zoo is not very near the British Museum or the Natural History Museum, and the great botanic garden is even further out, at Kew.[24] This was a royal garden. Loved by the parents of King George III, and adorned with a pagoda as part of the eighteenth-century passion for chinoiserie, it was put into the hands of Banks, who began to turn it into a botanic garden under King George's patronage, and who also supervised there the acclimatisation of the merino sheep, brought over from Spain despite the prohibition on their export, and subsequently sent to Australia to found the botany wool industry.[25] Banks had brought back a great number of specimens from his voyage, and he ensured that seeds and dried plants from later voyages were sent to Kew. He also arranged for collectors to be sent out, commissioned to get plants and seeds that could be cultivated there. His global and imperial vision set Kew on course to become a great centre for botanical science.[26] But as the King slid into senility, he fell out with Banks,[27] and his son, as Prince Regent and then as George IV, did not much care for Kew; nor did the next king, his brother William IV. The garden lost its scientific importance.

Then, in 1841, it was rescued and taken over by Parliament, as a national rather than a royal garden (though retaining the 'royal' in its official title); and William Hooker from Glasgow was appointed its director. Like the earlier Jussieu dynasty at the Jardin des Plantes, Hooker's family ran the garden for three generations, into the twentieth century, realising Banks' vision, and overseeing the building of enormous greenhouses that provided a model for the

Crystal Palace. William Hooker's son, Joseph,[28] himself went on a voyage to the Antarctic as a ship's surgeon, making himself the great authority on the plants of far southern latitudes; and then visited India, bringing back to Kew the rhododendrons from the Himalayas which became such a feature of Victorian gardens. He was also a great friend and prolific correspondent of Charles Darwin,[29] whose voyage had seemed a model for his own, and himself one of the great pioneers of evolutionary theory, experimenting at Kew. In Banks' day, most growing plants brought back on long sea voyages, through a series of climatic zones and exposed to salty spray or worse, had died; the introduction of miniature greenhouses, named Wardian cases after their inventor,[30] changed all that, and made possible some of Kew's triumphs in acclimatising plants and introducing them into pastures new, often in the colonies.

These successes, most notably with rubber and quinine from South America, but also with garden plants, built up Kew's reputation; and the steady flow of botanical literature, often beautifully illustrated by resident artists such as Walter Fitch,[31] sustained it as a great centre of botanical research and information. The herbarium grew steadily, and the greenhouses filled up. Magnificent publications such as *Curtis's Botanical Magazine* were associated it; and were indeed a kind of display. Pictures were displayed in the garden, notably in the pavilion donated by the great traveller Marianne North to show her work, and thus the splendours of exotic vegetation.[32] But the chief element of display about Kew was the garden itself. It was open to the public, at a cost of one penny. Joseph Hooker, a scientist of huge distinction, President of the Royal Society from 1873–8, was involved, in 1870–2, in a huge row with Acton Ayrton, First Commissioner of Works in Gladstone's government and thus responsible for parks, including Kew. He was brusque and dictatorial, a devotee of public economy, committed to cut public spending, and Kew (much more expensive than other parks, and with what seemed curious and idiosyncratic practices), with its wayward Director, looked a good target. Ayrton was offensive. A representation from the presidents of the major scientific societies and other distinguished men of science was sent to the Prime Minister, and Darwin's friend, John Lubbock MP, agreed to raise the matter in parliament. In the event, Hooker won; the populist Ayrton had underestimated what he took to be an effete naturalist with ideas above his station, and found himself reshuffled to another post – and losing his seat in the next election.

In fact, he was onto something: how were the functions of display and recreation common to public parks and gardens to be combined with those of a major scientific centre, and how were the expenses to be charged? The gardens had limited opening hours for the public which, through taxes, paid for their upkeep, and were costly in ways hard to understand for those outside the world of recondite science. It was probably inappropriate to include their maintenance in a government department otherwise concerned with land and buildings having no connection with scientific research. Moreover, Owen was busily campaigning for his new Natural History Museum to be built at public expense, and had not signed the memorandum of the great and the good in defence of Hooker's Kew.

There were herbaria in the British Museum going back to Sir Hans Sloane, its founder, and including much later botanical material: might not the science done at Kew be better transferred to the (still hypothetical) new museum, and the gardens turned into public space with exotic plants, tended by gardeners rather than scientists? In the event, Kew survived, strengthened by the controversy, which came at a moment when, in the wake of Prussian victory over France in 1870–1, scientific education and practice were suddenly becoming matters of importance to government, as connected with national wealth and power. And the wonderful arrays of plants in the flowerbeds, arboretum and greenhouses of Kew were available to give a taste of botany to ordinary people who could enjoy the display.

Like the Natural History Museum's, the botany at Kew, presented by the Hookers and their team, was establishment science: though visitors, able to wander where they would, could think their own thoughts more easily than inside a museum. But these were not the only museums established in London in this, their heyday. There had been specialised museums, like that at the College of Surgeons and that in Jermyn Street that went with the Geological Survey, where Huxley lectured on fossils. Then after 1851, and the success of that magnificent display at the Crystal Palace, land was bought in Kensington with the profits, and a museum of design set up, with Henry Cole as its patron. This was the ancestor of the Victoria and Albert Museum, and in the event, industrial design did not play much part in it. In 1862 a second 'Great Exhibition' was held in South Kensington, in what was agreed to be a boring building and in the shadow of the sudden death of Prince Albert: it neither made a profit nor aroused the excitement and imagination of 1851, particularly no doubt because it was the sequel to an act extremely hard to follow. But it did give a boost, and an opportunity for display, for industry and technology.

In 1876, in the wake of that Prussian victory and resultant boost for scientific education and training, an international loan exhibition of scientific apparatus was held in South Kensington. Associated with it (as with the 1851 exhibition) were lectures, and also daytime conferences in what seems a modern vein. The results were duly published. The twenty-five lectures were free, and therefore open to artisans, if not to unskilled workers, and were planned to bring the display as it were to life:[33]

> It became evident to those who were engaged in organising and arranging the loan collection of scientific apparatus, that its usefulness to the general public would be much increased, and the interests of science furthered, if explanations of the construction and uses of the various instruments could be given. Many of the exhibitors provided explanations at stated times of the instruments lent by them; but ... it was felt that it would be very desirable to have lectures on the classes of instruments and apparatus used for different purposes.... At this stage several scientific men came forward and generously offered their services in giving free lectures on the evenings when the collection was open to the public.

Some of the objects displayed were of historical importance: the first lecture, by Henry Roscoe, was on Dalton's apparatus and what he did with it; Playfair spoke on air and airs, with both the Magdeburg hemispheres and Boyle's air-pump among the exhibits; while both Tyndall and John Hall Gladstone described Davy's and Faraday's apparatus. Two of the lecturers were clergymen (both FRS); two lectures were given by Captain Davis of the Royal Navy on polar exploration; Lord Rosse spoke about telescopes; and the series concluded with a talk about weights and measures, timely in that the French metric system, now international, was becoming the standard in Britain for scientific purposes. The lecturer, here a poor prophet, noted that in the USA 'preparations are being made for adopting the metric system'. The chairman that night, the naval architect J. Scott Russell, referred to the exhibition where

> For the first time you see collected a large museum of instruments and apparatus which represent the great triumphs of human intellect and human science made during the last few centuries, and especially characteristic of the marvellous progress of science made during the century in which you and I have the good fortune to live and work.

And he wound up the proceedings with thanks to foreign exhibitors, and to the lecturers:

> I think it would be a great pity that such a collection should be dispersed. I think it would be a great pity that this should be the last lecture of this kind at which you and I are to have the pleasure of meeting each other.... We believe that the wonderful success which has attended these free lectures, and from the wonderful popularity which has been achieved by the lectures of profound and eminent men – as distinguished from mere professional lectures – profound and eminent men of science coming here and endeavouring to make plain to you the profoundest truths of their own special research, – I say it is a new future and a future which deserves great approbation on our part, and let us hope that the nucleus of a new museum to be entrusted I hope to the same able hands which have brought this together.

In this peroration, he addressed the audience as 'gentlemen', but elsewhere they were 'ladies and gentlemen' – we can probably infer that there were not very many women present, in audiences which various chairmen had complimented on their careful attention to what was said, and which were getting something not unlike the Discourses that their social superiors were hearing at the Royal Institution.

The thirteen Conferences ran from 16 May to 2 June 1876, and were modelled upon sectional meetings of the BAAS, but focused upon the loan collection. Some dealt with the historic pieces, but most were concerned with the latest science and technology. In this era, the acme of classical physics, exact quantitative work and standardisation of units (requiring precision and delicacy in apparatus) was crucial,[34] involving international rivalry especially with Germany. Spectroscopy, naval architecture, clocks, telescopes, submarine telegraphy, measurement in meteorology, terrestrial magnetism and sea-sounding were all discussed by the speakers. 'Mechanism' had by then through Babbage, Whewell and others, become a science, taught at Cambridge, where mechanics was applied to machinery.[35] The reports include the comments made by participants and by chairmen as well as the papers presented, and give us a good idea, fuller than in BAAS *Reports*, of what the sessions were like, and how in the context of a big exhibition, scientific enthusiasm was shared.

Russell had called for a new kind of museum, where apparatus and instruments old and new might be exhibited. Indeed, sandwiched between the Natural History Museum and the Imperial Institute (where colonial products were displayed) the Science Museum was founded,[36] and some unclaimed exhibits from 1876 found their way there. The site bought with the profits from 1851 was indeed being turned into 'Albertopolis',[37] a great cultural and educational centre, its museums comparable to those in Berlin or Munich, and to the Smithsonian Institution in Washington, DC.[38] From the very start, this museum, and similar ones elsewhere, faced the problem of how far they were historic collections of things used in important scientific work, and how far they should be providing hands-on experience to dazzle, enthuse and attract visitors into science. At the Natural History Museum, the collections behind the scenes are research material for people working on taxonomy, classifying animals and plants, and comparing the past and present distribution of organisms. In contrast, the Science Museum is not a centre for research in physics or chemistry: that goes on next door, in Imperial College. Similarly, the Victoria and Albert Museum is distinct from the nearby Royal College of Art, and is a centre of connoisseurship. Curators in both museums work on the provenance and authenticity of the exhibits, and on the best way to display them, but are not 'doing' art or science.

By the early twentieth century, there were thousands of what might be called science museums in Europe, North America and the colonies.[39] One of the great problems for major metropolitan museums is that they cannot display more than a small fraction of their holdings. This became a more acute problem in the twentieth century, as acquisitions came in, entertaining hands-on exhibits increased in popularity, and it became didactically fashionable to show fewer things and to have more description and guidance. In natural history, where, as Flower recommended, representative (or more spectacular) objects are on view and full collections accessible and necessary only to researchers, this may not matter so much; but when unique and priceless things cannot be shown to the public to whom they belong, being stored in a warehouse somewhere miles away, it is serious – especially when provincial museums would love to display them, often much closer to their original setting, where they could be appreciated properly. For provincial cities also had their museums, sometimes, like those in Exeter and in Oxford founded in connection with a visit from the BAAS. The Oxford one (scene of Huxley's confrontation with Wilberforce[40]) resulted from the separation of the science from the fine art in the Ashmolean

Museum (as subsequently happened with the British Museum in London), but most remained unspecialised, often including stuffed animals, minerals and fossils, archaeological finds, local history and model ships or railway engines in a more-or-less organised display. Science was a source of excitement.[41]

Such local museums could be more quirky, devoted as they might be to a local hero whose activities may not have been mainstream, and curated by people whose expertise could not extend across the whole realm of arts and sciences. Theirs was therefore less the established consensus view of things – they begin to take us into realms of doubt and uncertainty, into what might interest people other than science professors. The same may be true of the exhibitions and expositions which had begun before 1851, but which, following the huge success of the Crystal Palace,[42] became a feature of the second half of the nineteenth century.[43] They had to attract the public, display progress, boost the home city and country – being a great day out, worth travelling far to see, broadening visitors' perspectives, showing the latest inventions and discoveries and thus arousing enthusiasm for science and technology.[44]

Amongst the most celebrated of these were the 'Century of Progress' exhibition in Chicago in 1876 (marking confident recovery from the disastrous fire of 1871) and the Paris Exposition in 1900[45] where electricity was the great theme and the various industrial nations had their own pavilions in which to display their modernity. To accentuate this, a village from French colonial Africa, complete with 'natives', was also on display, demonstrating the benevolence of 'la mission civilatrice'. As well as electricity, phrenology was also available, and visitors could have their bumps measured, and other anthropometric data such as cephalic index, fingerprints, height, weight, eye colour, reaction time and dynamometer reading recorded, in the turn-of-the-century preoccupation with measurement and classification. The science might thus be rather softer than that in great museums, but the whole effect stunning.[46] The barriers around science had got taller as it became professional between the time of Banks and of Huxley; but the twentieth century opened with a great festival celebrating technology and science, even if ominously tinged with nationalism.

We may think of the twentieth century as the time when science and warfare came together, with deplorable slaughter as the outcome; with the nineteenth century as the time when science never did anyone any harm. This was not so: science had always been involved with the military[47] – the first President of the Royal Society, Lord Brouncker, had been an expert of the recoil of guns, Lavoisier worked on gunpowder, and lecturers at the Royal Institution regularly dealt with military explosives and warships. Voyages of discovery, like those of Captain Cook and of HMS *Beagle*, were the 'big science' of their day, getting a team of experts to distant locations: yielding exciting accounts, imperial claims, interesting objects, and trading possibilities. Military force is expensive, and needs constantly to be justified to taxpayers, but it lends itself very readily to displays. The Colt revolver had been one of the stars of the 1851 exhibition, and improvements in armouries meant that rifles soon replaced muskets as weapons for infantry. Shooting matches began, displaying their accuracy. William

Armstrong's rifled breech-loading big guns transformed the role of artillery, and more importantly for Britain, accompanied the disappearance of the great wooden sailing ships that had won the Battle of Trafalgar with ironclads, which had proved their worth in the American Civil War. Turner's sunset painting of the fighting Temeraire being towed by a steam tug to the breaker's yard symbolised the end of an era. The new Royal Navy, expanded in the effort to ensure that it was as powerful as the two next-largest put together, and crucial for the defence of a maritime empire, displayed itself in great reviews as both necessary and scientific:

> The navy was at the forefront of scientific and technological advance, provoking admiration and fascination. In its celebration the enthusiasm for modern technology was indistinguishable from expressions of national identity. Mock fights and spectacular searchlight displays were part of a distinctly modern imagery of power and technology.[48]

There were occasions on which such reviews were accompanied by displays of high technology, as when Charles Parsons' little *Turbinia* showed her paces, nipping hither and thither through the fleet, and convinced onlookers that the future lay with turbines rather than reciprocating engines.[49] These could even be contests, as earlier when in a tug of war, a naval screw steamer defeated one with paddles. Both these proved the greater efficiency of the new methods of propulsion; though, naturally, while the navy had to keep up-to-date and expense was no object, civilian paddle steamers, and reciprocating engines, did not disappear very soon, continuing in use well into the twentieth century.

Such public contests could also be a part of science. Louis Pasteur, whose research career began with careful sorting out of minutely different asymmetrical crystals in a laboratory, seems to have delighted in the public arena. In his public lectures he cultivated a kind of showmanship (as critics also said of Tyndall, who extended and disseminated Pasteur's work in Britain);[50] he also brought suspense and spectacle to his work on anthrax, the dire disease of sheep, and rabies. This had previously been the sphere of the quack.[51] Indeed, without medical training, he forced doctors (a notably closed and powerful profession) to take note of his experimental results by courting publicity, and building up his own myth.

This had happened earlier in England, with the original vaccination, against small pox. Vaccination in Britain had become compulsory as a public-health matter, and Wallace showed his alienation from the mainstream of scientists by opposing it along with other populists and libertarians.[52] Pasteur, whose work on germs was extended and popularised by Tyndall,[53] showed the possibility of vaccinating against other diseases; but he and his contemporaries also raised public outrage by the vivisection experiments they carried out. Huxley, who duly spoke up against what he saw as simply opponents of progress and enlightenment, had himself hated such research, and did not do it on conscious animals. However, his efforts to bring up-to-date physiology into Britain from France and

Germany entailed experiments on animals, sometimes without anaesthetic, and he thus threw his weight behind the campaign to continue vivisection, against the attacks Frances Power Cobbe and her formidable allies. Here science was indeed brought into the limelight, but not in a flattering way: dissectors could be portrayed as heartless torturers, coldly noting (if not enjoying) the squirms and squeals of their animal victims. This was not how scientists (especially in Britain) saw themselves, or wished to be seen – but rather as benefactors bringing medical advances. In the event, the scientific establishment (of which the 'plebian' Huxley was in later life a prominent member[54]) succeeded in getting a compromise where animal experiments were done only under Home Office licence. It is curious that experiments on human subjects, often medical students and sometimes self-experiment, but also hospital patients, continued with very little in the way of informed consent being required until into the second half of the twentieth century, when Nazi abuses were exposed. Not all publicity for science was good publicity, even in the Victorian age of science; understanding, private or public, had (and has) a moral aspect.

Ballooning, rail travel and crossing the oceans in great steamships became commonplace: disasters like balloon crashes,[55] the Tay Bridge collapse or the sinking of the *Titanic* were again poor publicity, but extremely instructive. They brought technical advance, often taken for granted, into the spotlight, reminded everyone of how dependent upon science life had become, and what gaps there were in understanding and mastery of nature. Sometimes also nature laid on a great and unexpected display of power, stimulating for science, like the eruption of Krakatoa,[56] probably the loudest event in history, and one of the most devastating until the earthquake and tsunami of December 2004 in the same region. The sunsets in Europe and North America became redder and more splendid, and meteorology and optics were among the gainers. The excitement generated by the disaster, the first to happen since remote regions had been linked to the rest of the world by electric telegraphs, was a stimulus to public understanding of science. For most people in Europe at least, the explanations had to be scientific rather than theological; but, of course, there was yet no tourism in Indonesia, and few Europeans had to answer questions like 'Why was I spared?' and 'Why was it my daughter who was killed?' Though few had visited these islands, except colonial officials and traders, there was intense curiosity in the nineteenth century about exotic regions. Wallace was among the travellers who had gone to what he called the Malay Archipelago, and his book about it became a classic: he was just one of the scientific travellers who form the subject of our next chapter.

8 Travel

One of the keys to happy and successful travel is to know where you are, and maybe where you are going.[1] This was not easy. The astronomer Edmond Halley was prominent among those seeking a way of finding longitude: magnetic variation might have provided the clue, but his charts and observations showed that it did not. Halley had been appointed captain of a naval ship, HMS *Paramore*, on a voyage into the South Atlantic to investigate terrestrial magnetism, in 1698–1701.[2] The voyage had been marred by a mutiny, which had forced a return to port so that his lieutenant could be put on trial. After this little difficulty, they returned to sea and met their objectives; but it became the policy of the Royal Navy that no outsider, however eminent in science, must again be allowed to command a ship. Astronomy, the study of the heavens, remained however the key to knowing where on earth you were. Instruments and navigation improved as Halley's Greenwich[3] and other great observatories built up data. Travellers' tales were no longer enough. The eighteenth century became a time of great scientific exploration, on land and water, as seafaring became more dependable and European empires grew and became less secretive; and the tradition carried forward into the nineteenth century. By then, travel books, often accounts of derring-do but frequently involving serious and accurate observation and description, making them of scientific importance, were becoming an extremely popular genre amongst stay-at-homes.[4] The great reviews made it their business to keep their readers well-informed about distant parts of the world. The exotic and sublime could be sampled vicariously.

The work of Linnaeus,[5] and the passion for rarities in gardens, made plant-hunting fashionable and important: pupils of Linnaeus undertook voyages and travels in search of plants for him to name, and many of them died, martyrs to science. Natural history and astronomy accompanied each other, bringing geography and Halley's terrestrial magnetism along with them, embodied in sailors who had acquired special skills and in civilian passengers, as survey voyages opened up new trade routes – naming and claiming coasts and islands as they went. Expeditions also went overland, as the Americas, Siberia, Australia and Africa were opened up – again by the military (notably the topographical engineers in the USA), by empire-builders, by hunters after big game, by gold-seekers, by missionaries like Dr Livingstone, and in the later nineteenth century by intrepid women like Isabella Bird, Mary Kingsley and Marianne North.

Specifically, scientific expeditions began with the transits of Venus in 1761 and 1769. Because the orbits of Venus and of the Earth are slightly inclined to one another, Venus only actually passes across the face of the Sun (rather than above or below it) at infrequent intervals: twice, eight years apart, and then not again for over 100 years. Halley perceived that if the transit were observed from places far apart on Earth, whose latitudes and longitudes were accurately known, then from this base line one could calculate how far away Venus and the Sun were from us. This would bring new precision into astronomy. He knew he would not live long enough to see this done; but after his death the astronomers of Europe, and in the colonies, resolved to make the observations. It was still impossible to fix the position of a ship at sea accurately. Latitude was no great problem, but only by carefully timing eclipses on a firm base on land could observatories determine their longitude relative to each other (fixing Greenwich as the agreed zero only happened in the nineteenth century). In 1761 there were many expeditions, some of which went to distant places only to encounter cloud and fog on the great day of the transit. Results were plentiful, but patchy and of varying accuracy. This transit could, however, be seen as a kind of dry run for 1769.

It was this one which brought James Cook upon the scene.[6] In war with Spain, in 1740 George Anson had taken a squadron into the Pacific, and captured the galleon taking treasure between the Philippines and Mexico. In most ways the voyage was a disaster, with huge loss of men and ships, but it brought South America and the South Sea into British consciousness. Other voyages followed, noting the Patagonian 'giants', but hastening on around the world because of Spanish hostility and dread of scurvy, which had killed so many of Anson's crews. But sailing round the world in 1766–8, HMS *Dolphin* discovered the island of Tahiti.[7] George Robertson, the ship's 'master' (a relatively junior officer) observed an eclipse and fixed the longitude; and, meanwhile, all enjoyed the fresh food of the island. Some sailors were at liberty to wander, where 'some of the Young Girls ventred over to the Liberty men, and our honnest hearted tars received them with great chearfulness, and made them some little presents'. An island with friendly natives and abundant provisions, on the other side of the world from Greenwich in a known position, seemed an ideal place to observe the transit: and Venus was not only a planet but also the presiding goddess of those welcoming girls, for whom nails and beads were the ideal presents.

Accordingly Cook, who had distinguished himself surveying the St Lawrence river in Canada in the run up to the capture of Quebec, was put in command of the ship to be sent, under the auspices of the Royal Society – and the suggestion that the geographer Alexander Dalrymple[8] should be made captain as Halley had been was firmly turned down. Cook chose a Whitby collier, capacious and manoeuvrable in tricky coastal waters, naming her HMS *Endeavour*. He was accompanied by an astronomer, armed with the new *Nautical Almanac* prepared by Nevil Maskelyne, now in charge at Greenwich, from which longitude could be found using precise observations of the Moon;[9] and by Banks and his entourage. The observations from Point Venus were successful; and the ship went on to New Zealand and New South Wales, surviving grounding and serious damage on the Great Barrier Reef. The voyage aroused immense interest. Here was adventure: noble savages, hostile warriors in New Zealand, naked men in Australia, animals and plants undreamed of, and temperate country previously unknown. The artists and surveyors left a wonderful record,[10] languages were learned and recorded,[11] and natural history collected.[12] Moreover, Cook had kept scurvy at bay, keeping his crew busy and making them eat salads: henceforward, voyages would be much less life-threatening, morale higher, crews healthier and more time available. The duties of ships' surgeons became less onerous, and their status higher: they had time for natural history if they wished.

The various journals kept on the *Endeavour* were passed to a literary clergyman, Dr John Hawksworth, to be put into suitably elegant form. He polished them (as engravers did with the drawings) or mangled them – but the book was a success, and the actual journals of Cook, Banks and others were not published until long after their deaths. They are more vivid, less literary, and more detailed than Hawkesworth judged the public would want. It is surprising that travel should have been perceived to need a populariser in the eighteenth century. Indeed Cook after his second voyage (when he carried a chronometer that made determining longitude much more straightforward[13]) insisted on publishing his own straightforward account, in what became the standard tradition; as did George Forster, the German naturalist who accompanied him (as assistant to his father, who had been forbidden to publish) when Banks dropped out.[14] Forster's writing aroused immense interest in his native Germany,[15] on the verge of the Romantic movement. This was the kind of science which the public could understand and appreciate.

In 1788 the First Fleet under Captain Arthur Phillip established the penal colony at Botany Bay, in fact moving immediately to nearby Sydney with its wonderful harbour. His journal was published,[16] anonymously edited by the publisher John Stockdale or one of his hacks, in a handsome volume 'embellished with fifty-five copper plates'. These included maps, charts and plates of plants and animals, including birds rather stiffly mounted on studio stumps. But early settlement produced enormous interest, and many handsome publications – clearly for the most part directed at the opulent end of the market. John White was the chief surgeon, and his journal[17] reveals the scientific traveller, wide-eyed, full of wonder, interested in everything in the new land which Banks, Solander and Cook had only been able to survey rather than explore. His illustrations revealed the wealth and strangeness of Australian natural history, and his own curiosity about its aboriginal inhabitants and their way of life. They had lived well off the land and its productions; but the new colonists, trying to preserve their precious seed-corn and animals, nearly starved. They could not find and eat witchetty grubs or other local delicacies, and waited anxiously for relief ships with supplies, and in the longer term for wheat, beef and mutton to flourish, giving them the diet they were used to. There was a real human drama in establishing a colony in the antipodes, especially with the involuntary exiles despatched there; and the animals and plants, once naturalised, had a devastating effect upon the indigenous people, as had happened in America. They found their country being turned into farms, ranches and sheep-walks.

Another First Fleet captain, and later Governor, John Hunter, also published his account – mostly of naval and human interest, but including plates of shells.[18] Then, in 1798–1802, David Collins, who had been Judge-Advocate there, brought out his journal, in two volumes; curiously, the book was translated by the German theologian Friedrich Schleiermacher in 1802.[19] The first volume contains many pictures of aborigines; the second has natural history engravings, notably of the wombat. By then, Matthew Flinders, who had already circumnavigated Tasmania, was undertaking with Banks' support the voyage around what he named 'Australia', looking in vain for navigable rivers or inlets into the interior. Flinders had been a midshipman with the irascible William Bligh,[20] trained by Cook, and himself trained John Franklin the explorer. Naval surveying was learned by this kind of apprenticeship, passing skills on by example rather than formal instruction.[21] Flinders was accompanied by Robert Brown, who was to become the doyen of British botanists.[22] Brown's appendix to the published journal, an essay on the flora of Australia, was a remarkable feat of generalisation. Not only had he made enormous collections, but he ensured that the study of plant geography became a central part of botany: description and classification were essential, but not enough. The collector must not just go for 'rarities' and 'nondescripts'; it could be as important to know that more-familiar species were also found in exotic locations. Study and comparison of floras was the way ahead.

Brown's work was stimulating to future travellers like Charles Darwin; but Brown's essay was hardly popular writing. Poor Flinders, his ship HMS *Investigator* falling to pieces, had taken charge of another, tiny, ship and sailed for home: but, landing at Mauritius, then under French control, he was put under house arrest as a suspected spy. The efforts of Banks to get him released were unavailing, and he was there for years. This was the more galling because Flinders had, despite the war, courteously greeted the French expedition to Australia under Captain Nicolas Baudin when they had met in the Great Australian Bight, and Baudin had also been (warily) welcomed in Sydney.[23] The French savants had fallen out with one another and the captain, but they produced some splendid pictures of what they saw.[24] They had gone to see what the British were up to, perhaps themselves to stake a claim in a new continent, and to invest their skills in describing the animal, vegetable and mineral resources – rather as an earlier French expedition had accompanied Napoleon to Egypt with very important results for Egyptology. But there was also the melancholy business of trying to find out what had happened to Jean Francois la Pérouse, the navigator whose ships had left Botany Bay in 1788 and disappeared. HMS *Endeavour* had nearly been lost on the Barrier Reef, and exploration was a service of danger like space travel in our day. The margins were narrow indeed: evidence of la Pérouse's shipwreck on a remote island was later found.

Because the light, the trees and the landscapes in Australia were very different from what artists, trained on Poussin and the picturesque, were used to, they found it difficult to paint what they saw, in that bright and savage land.[25] But public interest back home, and demand out there, saw painters mastering the new environment and demonstrating how different it was, and yet how colonists did their best to make it familiar. Sketching was part of a gentleman's education, and some of the naval men did it well. Among the convicts were talented artists and engravers, some of whom had used their talents in forgery back in Britain;[26] their output varied from the naive to the sophisticated, but they gradually caught the different quality of shade cast by eucalyptus trees from that of the oak and beech with which they had grown up. They also painted animals and birds, along with portraits of men and women, landscapes and townscapes; and there were also free settlers, such as John Cotton,[27] who made natural history paintings. It is difficult to get the aboriginals' perspective,[28] and while some colonists were kindly to individuals, probably few people in Australia or further afield were very interested in them as more than specimens. Certainly, their views would have seemed thoroughly pre-scientific to a generation that had grown up with the Enlightenment and Romantic movement.[29] Naturalists did however try to record local names for the creatures they encountered, though, as with 'kangaroo', there may have been occasional misunderstanding.

Latin America had been much longer known to Europeans than Australasia, but the Spanish and Portuguese governments, to whom it had been assigned by the Pope, were secretive about its vast resources. They regarded natural history and geography as state secrets, and feared that if the British, the French or the Dutch were to know too much, they might break into the trade there to a greater degree than the buccaneers and privateers were already doing. Indeed, the Dutch had, in the seventeenth century, tried to establish a colony in Brazil. Alexander von Humboldt, with Aimé Bonpland, getting out to Spanish America under French auspices, produced a sumptuously illustrated multi-volume masterpiece which not only described how the various sorts of flies on the Orinoco work shifts so that one is never free from their attentions, but also included analyses of the Mexican economy, and accounts of how plant life changes with altitude. Humboldt's, reprinted in more accessible form and translated into English,[30] was a 'personal narrative', where his feelings and reactions were there with the facts. He also published essays, seeking to convey how views or aspects of nature are enriched by scientific training.[31] His writings, translated by learned women, were available in competing cheap editions by mid-century; he was one of the first to think globally.

He called on Thomas Jefferson on his way home, when Jefferson had just bought the West from the French, in the 'Louisiana Purchase'. This was largely unknown; and expeditions, often behind pioneering settlers building their little houses in the big woods and on the prairies, were sent out, to report on the country and then (when California had been wrested from Mexico) to find the best route for a transcontinental railway – published by Congress in a sumptuous series of *Pacific Railroad Reports*. Charles Mason and Jeremiah Dixon, surveyors trained on expeditions to observe the transit of Venus, had perhaps begun this tradition in 1763-7 with their famous line between Maryland and Pennsylvania, later the front in the Civil War. There were men of science in the USA engaged in abstruse studies such as electromagnetism or thermodynamics, but the main emphasis (not surprisingly, given that the frontier closed so recently that my grandmother could remember it) was on 'Humboldtian' geography and natural history. There was an important American voyage into the Antarctic and the Pacific, under Charles Wilkes, in 1839-40, which produced its store of pictures and specimens;[32] and the 'opening up' of Japan under Commodore Matthew Perry in 1852-4 was clearly a very important event, but not directly scientific.[33] The plants and animals of Japan had become known in the west earlier through the Dutch trading base or 'factory' at Nagasaki, and the annual pilgrimage to bring tribute to the Shogun at Tokyo, but could now be studied at first hand. Things Japanese became a craze in the west.

Scientific voyages were thus no British monopoly; but, because a trading island nation depended upon merchant shipping and therefore good charts, and because the Royal Navy was so large, and there was no major war until the 1850s when fleets were in action against Russia around the Crimea and in the Baltic, hydrography was a particularly British activity.[34] The suppression of the slave trade required squadrons to patrol the coasts of Africa, dhow chasing in the Indian Ocean,[35] and intercepting slave ships bound for the USA or Brazil in the Atlantic. This activity was very popular at home, especially among evangelicals who at this time had a strong social conscience; but sailors were less enthusiastic about hanging about off feverish coasts, though getting some satisfaction from putting an end to a vile business. Prominent captains were James Tuckey, with experience of Australian surveying, who died with many others in his party exploring the Congo,[36] and William Owen, whose charting of the coasts of Africa on board HMS *Leven* was very important.[37] He was particularly impressed with Madagascar, where he signed a treaty with King Radema who, in return for arms with which to suppress independent chiefs, agreed to suppress the slave trade. In Owen's wake came missionaries from the London Missionary Society, who in subsequent years maintained interest in the enormous island, its people and its extraordinary natural history.[38] The Navy and spreading the gospel together made 'Humboldtian' science highly palatable to Victorian readers.

Another captain who died on duty was Henry Foster, a man whose scientific reputation had secured him command. He had received the Royal Society's Copley Medal in 1827.[39] In the navy, as in engineering at the time, there was wide suspicion among practical men of book learning, and Foster was envied. Fixing places on the two sides of the isthmus of Panama using chronometers, he was drowned in a canoe accident on the way back to the ship. Meanwhile, HMS *Beagle* was sent to survey around Cape Horn. In these desolate and dangerous regions, lonely and depressed, her captain Pringle Stokes shot himself; on her return to England, Robert FitzRoy was appointed to the command. A companion, as Banks had been for Cook, seemed a good idea; Charles Darwin was the man, making the ship and her voyage (which, after all, was just one of a long series of surveys) one of the most celebrated ever.[40] Darwin brought to the voyage an enthusiasm from reading Humboldt for tropical flora, and for seeing the world; his family was associated with opposition to slavery; and he had been informally trained in botany and geology, one-to-one with Cambridge professors.

The voyage, where the years in South America were probably more significant than the stopovers at the Galapagos Islands and in Australia (instructive though those were), turned the young would-be country vicar into a scientist: though, at first, the published fruits, the narrative and the big volumes of zoological descriptions by a team he assembled, marked him as a 'traveller'. His study of barnacles was important in getting him recognised as a real biologist, entitled to be taken seriously. But, just as Banks had become President of the Royal Society, so did Edward Sabine, whose wife translated Humboldt; Thomas Huxley, who had been assistant surgeon on HMS *Rattlesnake* surveying Australia and New Guinea;[41] and Joseph Hooker, who had sailed also as a surgeon to the Antarctic with James Clark Ross – becoming the great authority on plants of high southern latitudes, and Director of Kew. It was Sabine who presented the Copley medal (rather unhappily) to Darwin.[42] Ross' voyage, which took him into the Antarctic sea which bears his name, was primarily for investigating the Earth's magnetism, as part of an international programme. The ships were floating magnetic observatories.[43] John Herschel, whose expedition to South Africa to observe the southern stars had been private, became a heroic and immensely respected figure as a result, the sound man to consult on scientific questions.[44] Like the scientists, the captains on these voyages (if they survived) mostly did well in their careers, and finished up as admirals.

HMS *Beagle* made another survey voyage, around Australia, where Darwin's old shipmate, John Lort Stokes, now captain, named a harbour in the north after him.[45] She was filling in detail on the surveys begun by Flinders, and carried on by Phillip Parker King, who had sailed in the tiny HMS *Mermaid*, of 84 tons, built of teak in India, with a crew of nineteen, including an aboriginal interpreter.[46] Both captains included much natural history and anthropology as well as geography in their books. The public seems to have delighted in the plain workaday prose of naval men, and their stories made momentous by science. As people became more prosperous, so tourism began where scientific travellers had gone; but the distinction between travel and tourism remained important to travellers.[47]

Naval surveys are, however, like going for a cruise: you learn a lot about coasts and islands, but not about what lies within the continents. Tuckey's team picked up some central African botany from what the Zaire river brought down; King and Stokes followed up rivers in their boats as far as possible; while Darwin left HMS *Beagle* for long periods to explore inland, while the crew were doing detailed charting. Humboldt's expedition had been along rivers; and, in Canada, rivers were the key to opening up the country.[48] In the Canadian arctic, John Franklin made an amazing journey down the Coppermine river to the polar sea, on which he navigated in birchbark canoes.[49] On the way back, the party nearly starved to death. John Richardson, the doctor, recognised human bones in a stew made by one of the guides, which explained the disappearance of another. Having shot the offender and thrown away the dinner, they were reduced to eating their boots before being rescued by local Indians. They were doing more science than Thomas Hearne, a fur-trader who had been there before; or Alexander Mackenzie, who in 1794 had reached the Pacific Ocean.[50] He had nearly met George Vancouver, who had been sent to see the Spaniards off the island that now bears his name, and to chart the coasts of what is now British Columbia and Alaska.[51] Franklin later went right down the Mackenzie river, in an attempt to link up with Beechey coming through the Behring Straits.[52] These surveys put paid to the seductive idea that there was a practicable north-west passage, a polar route to the far east, but maintained British imperial claims in the face of the Spanish, the Russians and later the Americans. In the end, Franklin returned to the arctic in 1845 with the ships that Ross had used, HMS *Erebus* and HMS *Terror*, in a bid to get from east to west along the coast he had explored earlier. The party filled in the missing links on the map; but the ships were trapped and crushed in ice, and the crew all died, on board or trying to make their way overland to civilisation. The 'search for Franklin', with private and naval expeditions,[53] led to the mapping of these desolate territories and much knowledge of climate, auroras and natural history, as well as geography, amid great public interest in what was really rather like the space programme a century later.

Even these overland journeys had often involved naval men, though they were not used to tramping across country; and, in 1849, Herschel had edited for the Admiralty, which seemed to have a monopoly in exploration, a *Manual of Scientific Enquiry*.[54] This has a chapter on Geology, which now makes the book famous; Herschel had hoped that Adam Sedgwick would write it, and when he refused, Darwin was the second choice. Other authors included George Airy, the Astronomer Royal, and Henry de la Beche, Director of the Geological Survey, Sabine, Beechey, Whewell, Owen, William Hooker, James Prichard the eminent anthropologist, Alexander Bryson, who was shortly to become Inspector General of Hospitals and Fleets, and G.R. Porter, a prominent statistician: it was a formidable team. Prizes were offered to ships' surgeons who submitted the best journals after a voyage. Like Faraday's *Chemical Manipulation*,[55] the book has dated remarkably little: much more is known about geology and meteorology, but the recommendations for scientific travellers would still be valuable for the inquisitive, anxious to find out about distant places. The authors did not expect prior knowledge, conveyed enthusiasm, and provide us with a good picture of what descriptive science was like about 1850. Cook and Banks at Rio, and Flinders at Mauritius, had been strongly suspected of spying, and this book indicates how such fears were not unreasonable. An intelligent and well-prepared naval officer would notice all sorts of things that secretive governments might like to keep to themselves, though he would never think that this was espionage.

Cook had disproved the notion that there was an enormous southern continent, to balance the land in the northern hemisphere, and inaugurated surveys to assist merchant shipping; 100 years later, as undersea telegraph cables were linking the continents, the Royal Society and the Royal Navy sponsored research into the ocean depths, with the voyage of HMS *Challenger* around the world in 1873–7.[56] An hypothesis being tested was that of Edward Forbes, that life in the deep sea, in the darkness beyond the reach of any sunlight, would be impossible. Cable-laying ships had cast doubt upon this, and *Challenger* disproved it completely. This generated an enormous mass of scientific collections and data – the ship was even fitted out with a chemical laboratory – but there were also accessible writings, making the voyage intelligible, from the marine biologist Charles Wyville Thomson.[57] He also played a major role in publishing the formal results in forty volumes of text, and six of atlas, between 1880 and 1895, in a major feat of international organisation of research, completed by John Murray after Thomson's death. The captain was G.J. Nares, who was recalled from the *Challenger* at Hong Kong to take command of an expedition to the Arctic, setting sail in May 1875, watched by cheering crowds.

Indeed, there was great public excitement; and, like Ross' to the Antarctic, the voyage's scientific objectives included magnetic observations. Great care was taken that the two ships should not be together in high latitudes, so that both would not be frozen in and crunched. After wintering, a party got to within 400 miles of the North Pole, but contracted scurvy, and had to be rescued by their shipmates, who had been mapping 300 miles of icebound coastline. National prestige, easier to communicate to the public, and science formed a somewhat uneasy mixture in these polar expeditions, which by the twentieth century involved the Antarctic, where Robert Falcon Scott's expedition in particular had scientific objectives. The work of Edward Wilson, the naturalist, was particularly noteworthy,[58] but the public could appreciate the heroism and grit of the heroic explorers much more readily than most of the science involved. And even well into the twentieth century, there were more congenial parts of the globe still unmapped.

While he had climbed upon the eternal snows on Chimborazo, Humboldt's overland travels had been on the steamy Orinoco, in Peru, and in Mexico: he had not been allowed to enter Brazil, Portuguese territory. Darwin had been in Argentina and Chile, but the Amazon region remained little-known to science. Two men who became Darwinian disciples changed all that. Alfred Russel Wallace and Henry Walter Bates, stimulated by reading Humboldt's and Darwin's travels, went together to the Amazon in 1848 to make and sell collections of natural history. A thousand miles up river, they parted in 1850 and went their separate ways, Bates remaining in Brazil until 1859, when he described himself as 'an oldish yellow-faced man in big whiskers'. He had become a great authority on butterflies, especially struck by how some species from genera palatable to birds have come to look like quite different and bad-tasting species. He concluded that this 'mimicry' was the result of evolution by natural selection: individuals which had looked less good to eat had been spared, and therefore left descendents, and in each generation those most like the unpalatable butterflies had been spared, so that gradually the whole species had come to resemble the inedible one. On his return, he read papers about these butterflies in London in 1860, and corresponded with a delighted Darwin. Bates estimated the total return of his eleven years on the Amazon at £800. Darwin recommended his travel book to Murray, publisher of the *Origin of Species*. When it was published, Darwin praised it as 'the best book of Natural History Travels ever published in England'. The edition of 1,250 copies soon sold out. Murray persuaded Bates to abridge it, leaving out much of the science, for a second edition. He was then instrumental in getting Bates a job, administering the Royal Geographical Society. In 1892, he published another full edition: readers of Bates picked up Darwinism painlessly.[59]

Wallace had been joined by his brother, who then died of yellow fever. He set off for England with his collections, but the ship caught fire, the collections were (like Raffles') destroyed, and only after ten days in open boats were the passengers and crew rescued. Wallace went to work classifying the specimens he had sent home earlier, met leading men of science, and wrote his *Travels on the Amazon and Rio Negro*, an excellent travel book. He came to see the loss of his main collection as providential, because instead of spending years in a museum classifying, he set forth in 1854 upon another journey, to the Malay archipelago. There was a considerable literature, going back to Raffles and before, on the area: a voyage to the Moluccas and New Guinea by Thomas Forrest;[60] a history of Sumatra by William Marsden;[61] an account of an embassy and then a *Descriptive Dictionary* by John Crawfurd[62] – all containing some science. But Wallace was not a colonial official, or a merchant: he was a man of science, remarkably free of casual Victorian racism,[63] living happily among Dyaks, despite their head-hunting reputation. He was particularly interested (as were contemporary stay-at-homes) in orang-utans, wild men of the woods to Malays, and *Homo sylvestris* to Linnaeus, who had therefore put them very near us in the scale of nature. Our ape relatives could not but interest an evolutionist. But Wallace worked on insects too, and his book, *The Malay Archipelago*,[64] blended natural history and anecdotes of travel with powerful generalisations. He was, like Bates, what was called a 'philosophical naturalist', reflective about what he saw.

He saw that the fauna and flora of Bali and Lombok were very different, although you can see one from the other: Bali is Asiatic, Lombok Australian, in character. The deep strait between them separated plants and animals that had come from the two neighbouring continents. Such boundaries are now named 'Wallace Lines', after him. Then, on the Moluccas, recovering from a bout of fever in 1858, he hit upon the idea of evolution by natural selection, and wrote his famous letter to Darwin which entitles him to the title of co-discoverer, though Darwin had been working on the idea for years, and Wallace modestly dedicated his book to him, referring always to their theory as 'Darwinism'.[65] When he got back from Malaysia, after eight years away, he had collected 125,000 specimens of natural history, assisted by various Malays. Wallace's open-minded and clear writing made his one of the great travel books, wonderfully evocative and readable, making understandable the attraction of science.

He was no dry-as-dust professor. Indeed, what struck his contemporaries was his readiness to go where his curiosity took him, into phrenology, spiritualism, theism, socialism, and opposition to vivisection and vaccination[66] – causes that to many looked opposed to scientific progress. But he was read widely, and his readers could not but realise that science is a broad church rather than the narrow orthodoxy that some experts might prefer.

Darwin also much delighted in Thomas Belt's book about Nicaragua;[67] sending a copy to a German colleague, he described it as 'the best book of natural history travels ever published'. Darwin liked his authors to be curious and generalise rather than just describe; and Belt was a convinced evolutionist, thinking things through. For their frontispieces, Darwin's favourites chose striking engravings: Wallace had dyaks attacking an orang-utan, Bates showed himself beset by toucans, and Belt's shows him ambushing a jaguar. It is good to get pictures of naturalists, like other animals, in their habitats. Belt discussed everything: slavery, reasoning among ants, 'mimicry' in butterflies, geographical distribution of animals and past geological changes, and how bright coloured animals might have evolved – writing of a macaw that:[68]

> This gaudy-coloured and noisy bird seems to proclaim aloud that it fears no foe. Its formidable beak protects it from every danger, for no hawk or predatory mammal dares attack a bird so strongly armed. Hence the need for concealment does not exist, and sexual selection has had no check in developing the brightest and most conspicuous colours. If such a bird was not able to defend itself from all foes, its bright colours would attract them; its bright colours direct them to it own destruction.

All this came in the context of a narrative of travel, including visits to goldmines; the science, trained and organised common sense, comes naturally and easily out of the story.

Also from Latin America, this time Argentina, came the wonderfully evocative writings of William Henry Hudson. He achieved fame and a glowing review in *Nature* by Wallace, with his book about La Plata, where he had grown up[69] – regarding Argentina as 'the purple land that England lost' in never adding it to the British Empire. Seriously ill in his teens, an invalid, and a sadder and a wiser man after reading Darwin, he left Argentina and an idyllic childhood there he could never forget. Coming to England in 1869, and looking on it with eyes familiar with very different landscapes, he wrote a wonderfully fresh book on the birds of smoky London; and other writings which particularly appealed to readers in the great cities of late-Victorian and Edwardian Britain, nostalgic (like Hudson) for the countryside world they had lost.[70] He was one of the founders of the Society for the Protection of Birds (later 'Royal'), which helped to make Britons a nation of bird-watchers and lovers. Hudson brought to nature the clear-sighted vision of the old-fashioned naturalist, sharpened by then with binoculars that made careful observation of behaviour possible, and removed the need to kill the thing one loved. Travel worked both ways.

Sir William Hamilton, whose travels took him only to Italy where he studied Vesuvius,[71] was, with Banks, a member of the Society of Dilettanti. They were a group interested not only in works of art, but also in natural history, and their name had none of the negative connotations of amateurishness that it has today. The Society had, in 1764–5, sent Richard Chandler further east in the Mediterranean, to Asia Minor, to lead an expedition looking primarily at antiquities, marking a beginning of the Biblical archaeology which so fascinated Victorians. Chandler reported on earthquakes, petrifying streams, suffocating caverns and plague, and sought in vain for any remains of the Temple of Diana at Ephesus.[72] Soon, attention shifted further round the Mediterranean, via Arabia and Egypt, and Banks was among those who formed the African Association, interested in opening up the continent.[73] Among the great geographical problems were the source of the Nile, and the course of the Niger; and intrepid adventurers were dispatched, often disguised as Moslems, to investigate. These travels were also associated with the suppression of the slave trade, and culminated in the founding of the Royal Geographical Society, long to be ruled over by Roderick Murchison. The exploration of Africa, with such great names as Andrew Smith,[74] Mungo Park, Speke, Grant, Burton, David Livingstone,[75] Henry Morton Stanley and others aroused great enthusiasm and involved both natural history and practices akin to navigation and surveying, as the travellers plotted their paths across spaces previously blank on the map. With Stanley's search for Livingstone, or later to rescue Emin Pasha,[76] human drama was added to the derring-do and careful or ruthless planning. In this era before two cultures, such geographical science was simply part of the whole enterprise of exploration, evangelising and abolition of slavery: there was no problem about its public understanding, or indeed support.

Great lakes, and enormous snow-covered mountains, were reported by the astonished travellers, along with a wealth of natural history. Especially notable were the reports of the anthropoid apes: Wallace had observed orang-utans, but the observations on gorillas published in 1861 by Paul Belloni du Chaillu, a French-born American, caused a sensation and were dismissed by many as mere traveller's tales. Since the *Origin of Species* had just come out, the topic of our relationship to the apes was highly loaded. To us, it is sad that natural history required considerable slaughter, euphemistically called collecting; but, for purposes of dissection and classification, it was necessary to kill a number of specimens. Travellers also needed to eat, so stalking and shooting were an important part of exploration. But a new breed of wealthy men who delighted in killing large animals (their cousins back home contented themselves with destroying deer, foxes and grouse) appeared: big game hunters. Because successful hunting demands knowledge of the character, behaviour and anatomy of one's prey, it has often gone with deep knowledge of natural history – at least on the part of trackers, guides, gillies and gamekeepers bringing their employers within range. But there were hunters whose pictures and descriptions of animals and their habitats raised the genre of exiting tales of hunting formidable animals into real natural history, as with Cornwallis Harris in South Africa in the 1830s.[77] The splendid lithographs here and in Andrew Smith's work (which had been supported by a British government grant) finally put to rest Dürer's splendid rhinoceros, with its Renaissance-style armour and extra horn between its shoulders.

Meanwhile, in India, ancient maps were being superseded by the great project of a full trigonometrical survey, completed in 1841 under the direction of Sir George Everest, who duly got the mountain named after him. This was one of the ways in which India functioned as a laboratory where experiments could be made which, if successful, could be repeated in Britain. The Survey, especially as anxiety grew about Russian intentions in Central Asia, began sending brave Indians into and beyond frontier territories up into and beyond the Himalayas; and some European travellers went there too, notably William Moorcroft, one of the very first professional veterinary surgeons, who was in search of good horses for the army.[78] There were also important travellers visiting from Europe, like Joseph Hooker, as well as soldiers and administrators who made natural history in its animal, vegetable and mineral branches a speciality. Among geologists, William Henry Pratt, a Cambridge mathematician made Archdeacon of Calcutta in 1850, and dying in India in 1871, made a name for himself. The Calcutta botanic garden, and the Governor-general's menagerie at Cawnpore, were centres of science; and the ultimately successful attempts to establish tea plantations in India were a good example of acclimatisation. The Indian exhibits at the Crystal Palace, and then at the Imperial Institute, attracted attention. Here again there was little difficulty in communicating the science involved; especially when rapid communication with India, and then with North America and Australia, was achieved through telegraph cables.

Most travellers were men, but especially those working in India were sometimes accompanied by their wives, who might be keen observers of humankind, natural history and geography. Thus Honoria Lawrence accompanied her famous husband Henry in his tours of duty in India, giving birth there to a large family, and writing diaries and letters which give a vivid picture of what life looked like to a keen-eyed woman, often the first western woman that locals had seen.[79] By the later nineteenth century, there were independent women travellers, including Isabella Bird, Mary Kingsley and Marianne North, whose travels from 1871 after her MP father died took her around the world, depicting the plants she loved in their natural settings.[80] Her lush paintings now fill the gallery she presented at Kew, named after her and opened in 1882, the year of Charles Darwin's death (he had told her she must visit Australia; and she had hoped he would have been able to open it). She knew all the right people, notably Joseph Hooker and other British men of science, including Frances Galton, whose *Art of Travel* is a wonderful compendium;[81] but also met the Emperor of Brazil, and in the USA met the Grants in the White House, and the Adams family at Quincy. Hers were not straightforward botanical illustrations, though she was a botanist and they are not just decorative flower paintings: they show the place of enthusiasm and imagination in science, and notably in its communication, to which we now turn.

9 Imagining

Science, in the form of pictures, exhibitions or travel books, might be easy to engage with, and ripe for public understanding. There were naturally also all sorts of possibilities for misunderstanding, reinforcing nationalism and stereotypes, which could also go with public enthusiasm. But much science is arcane, indeed rather dull: chemical analyses, astronomical observations, classifying plants, tabulating magnetic and meteorological readings do not lift the spirit. Something startling, some anomaly, may present itself to the trained mind doing these things, but most of the time this 'normal science' requires skill, discipline, trustworthiness, and other such solid and mundane virtues. Careful observation and accurate reporting, scrupulosity indeed, are an important part of what a scientific training is about. Mastering technical terms, formulae and equations, statistics, manipulation, technique: these things are the grammar of science, necessary for those who want to pursue it. Those who don't master these things are all too liable to be carried away by quack treatments, nostrums, panics and dogmas. Public understanding of science is always going to have to face the problem that the natural sciences are 'unnatural': carefully controlled laboratory experiments, lengthy statistical investigations, graphical print-outs, or microscopic taxonomic work are recondite and forbidding to most of us.[1] Anything worthwhile has this element of training about it; everyone knows that real 'effortless superiority' is a chimera. Science is not distinct from other difficult activities, like practising law, or playing sports at international level. Natural history such as bird-watching can provide the equivalent of playing games for fun rather than for a living, but isn't what a lot of science, particularly the expensive kinds, is like.

In early Victorian Oxford and Cambridge, gentlemen received a liberal education. This was based upon Classics and mathematics, and the idea was that everyone would share this common basis of civility – knowing about the past, and about exact reasoning. There were professors of theology, law, medicine, chemistry, astronomy and modern history, but these subjects were not part of the undergraduate syllabus. Professional subjects like law or medicine, or theology enough for ordination, would be learned later: this way of entering chemistry was indeed advocated by Davy, who himself lacked any higher education. Elementary mathematics, the first principles of general physics, and[2]

> Latin and Greek among the dead languages, and French among the modern languages, are necessary; and, as the most important after French, German and Italian. In natural history and in literature, what belongs to a liberal education, such as that of our universities, is all that is required; indeed a young man who has performed the ordinary course of college studies, which are supposed fitted for common life and for refined society, has all the preliminary knowledge necessary to commence the study of chemistry.

Professional subjects were not appropriate to common life or refined society. Men would have to learn these modern languages out of school or university – maybe like Byron from girlfriends. A man without a liberal education might, in Davy's view, 'be a good practical chemist', but 'he never can become a great chemical philosopher'.

The problem that faced Davy and his contemporaries was how to get chemical or natural philosophy across to their publics, who were not going to become analysts or engineers, without having to go too far into the professional, practical, sphere. Especially as medical qualifications came after 1815 to require formal courses and examinations in sciences, and then degrees in chemistry, physics and biology were introduced, these graduates might seem to ladies and gentleman to be illiberal and unpolished, though well-trained, technicians. Newton, Davy, Faraday and Tyndall were charismatic figures, relishing the title of philosopher; Jane Marcet or Mary Somerville could not, as women, be mere practical men; but Brande or Parkes might seem a bit unphilosophical, people to whom the new word 'scientist' might be more appropriate – especially as Whewell coined it by analogy with 'artist', which in 1833 might easily mean 'craftsman' (despite the efforts of Joshua Reynolds and his associates in the Royal Academy to become fine and genteel). Thus the scientist might look like an expert craftsman, a kind of superior plumber or glassblower. When Oxford and Cambridge began science degrees, it was suggested that ancient Greek might not be necessary for candidates (despite Davy, and words like 'oxygen') as part of a new BSc degree. But amid fears that this label would indicate training for a practical man, rather than education for a gentleman, the proposal was dropped and the degree remained a BA like the traditional ones and subject to their requirements.

What distinguished the philosopher from the technician was the shaping spirit of imagination. S.T. Coleridge, whose opposition to men of science calling themselves philosophers had prompted Whewell to invent his new word, had earlier recorded his admiration for chemists like Davy:[3]

> Thus, as 'the lunatic, the lover, and the poet' suggest each other to Shakspeare's Theseus, as soon as his thoughts present to him the ONE FORM, of which they are but varieties; so water and flame, the diamond, the charcoal, and the mantling champagne, with its ebullient sparkles, are convoked and fraternized by the theory of the chemist. This is, in truth, the first charm of chemistry, and the secret of the almost universal interest excited by its dis-

coveries. The serious complacency which is afforded by the sense of truth, utility, permanence, and progression, blends with and ennobles the exhilarating surprise and the pleasurable sting of curiosity, which accompany the propounding and the solving of an Enigma. It is the sense of a principle of connection given by the mind, and sanctioned by the correspondency of nature. Hence the strong hold which in all ages chemistry has had on the imagination. If in SHAKSPEARE we find nature idealized into poetry, through the creative power of a profound yet observant meditation, so through the meditative observation of a DAVY, a WOOLLASTON, or a HATCHETT ... we find poetry, as it were, substantiated and realized in nature: yea, nature itself disclosed to us ... as at once the poet and the poem!

To astronomers, chemistry might seem like cookery; but, to its devotees and their admirers, it was a science showing the connectedness of things, the underlying simplicity grasped by the mind through profound imagination. The world was not the great clock that Newtonians had supposed it to be: it was a poem.

We think of imagination as the power of making things up.[4] It lies behind fairy tales rather than chemistry, though in our day one might see it in cosmology: and if tough-minded scientists called that 'imaginative', they would not mean to be polite. Imagination is the realm of fiction. And yet we know that fact can be stranger than fiction, and that situations and roles can be imagined in drama and stories which can help us to understand and take action in our real world. For Coleridge, imagination was different from what he called 'fancy'. Fancy was fanciful, a world of similes, of images yoked together for effect but having no real connection: light verse, puns, jokes that don't go deep, word-play and the tired 'poetic diction' of the eighteenth century. Imagination was the plastic power that led us to perceive the true links between things, to go below the shell and surface of phenomena to the underlying reality that shallow minds miss. Thus the 'Ancient Mariner' is imaginary, but also imaginative literature, full of allegory and suggestion, so that, like the wedding guest, having heard or read it, we cannot forget it.[5] Prose seemed in the generation after Erasmus Darwin to be the medium for science: indeed, science seemed prosaic to many – but the right, dynamical, kind of science was, for Coleridge and his friends, creative, imaginative and poetic. William Wordsworth, who in 1800 opposed Poetry to Science rather than to Prose, had also referred to 'the pleasure which the mind derives from the perception of similitude in dissimilitude': and it may have been this in Davy's work, as well as their acquaintance, which led him to appreciate that the real man of science, like the real poet, must be imaginative. In 1815, distinguishing it from fancy, he wrote of imagination, referring like Coleridge to *Midsummer-Night's Dream*:[6]

> That Faculty of which the poet is 'all compact;' he whose eye glances from

earth to heaven, whose spiritual attributes body forth what his pen is prompt in turning to shape; ... Imagination, in the sense of the word as giving title to a Class of the following Poems, has no reference to images that are merely a faithful copy, existing in the mind, of absent external objects; but is a word of higher import, denoting operations of the mind upon those objects, and processes of creation or of composition, governed by certain fixed laws.

This is the world of powerful metaphors, found out by deep thinking, which like scientific theories and models genuinely cast light upon the way the world is.

We cannot really doubt that natural philosophers have always made imaginative leaps in the dark, seeking afterwards to test their conjectures by observation, mathematical analysis, experiment, or search for coherence and consistency. But that was not how science was presented in the years around 1800. Anxious to avoid the 'systems' of eighteenth-century philosophes, with wide-ranging and untestable world-views like Voltaire's or Erasmus Darwin's that impinged closely on religion and politics, men of science in France as in Britain turned to Bacon as the advocate of cautious induction.

In that second scientific revolution, as specialised experts began to replace generalists and science became more exact and forbidding, John Herschel's *Preliminary Discourse*[7] presented a sophisticated inductive method, with deference to Bacon, but with awareness of how Newton's mathematical way departed from Bacon's recommendations. His book stimulated both John Stuart Mill, whose view of science in his *Logic*[8] was thoroughly inductive; and William Whewell, whose *History* and then *Philosophy of the Inductive Sciences* presented a different view, where the thinker must hit upon the right end of the stick in order to be able to collect relevant facts, superinducing theory upon observation.[9] There seems no doubt that Whewell was closer to what his great contemporaries like Michael Faraday and Charles Darwin were doing: though Whewell was appalled by the *Origin of Species*, it was the offspring of his method of deduction and testing. Contemporaries were puzzled by the way that Faraday kept making astonishing discoveries: because of his lack of formal mathematics, they thought of him as an unschooled experimental genius – but we cannot doubt his imaginative mind, looking for deep resemblances, and forces underlying matter.

In France, Auguste Comte perceived human knowledge, in society and in the individual, as going through three stages, according to his Law of human progress:[10]

The law is this: – that each of our leading conceptions, – each branch of our knowledge, – passes successively through three different theoretical conditions; the Theological, or fictitious; the Metaphysical, or abstract; and the Scientific, or positive. In other words, the human mind, by its nature, employs in its progress three methods of philosophizing, the character of which is essentially different, and even radically opposed. . . . The first is the necessary point of departure of the human understanding; and the third its fixed and definitive state. The second is merely a state of transition.

The world was now arriving at this third stage, becoming aware that the only way of arriving at truth was by scientific methods: thus everyone in their education should be hustled through the necessary preliminary stages into scientific enlightenment. This might allow for some scientific imagination – though, despite his mathematical training, Comte's view of method seems close to Bacon's and Mill's; but, in general, imagination would go with fictions, to be put away along with other childish things as we, and humanity, grew up. It seems a rather chilly vision.

Indeed, that struck Comte, who invented, to go with this austere world-view, a Religion of Humanity,[11] in which altruism (a Positivist word) was emphasised, a Raphael Madonna (and then Comte's girlfriend, Clotilde de Vaux) taken as a symbol of humanity, and a ritual very like Roman Catholicism introduced. Positivist churches were founded, in Britain as in France, where with due ceremonial, a catechism, and a special calendar were introduced.[12] This began on 1 January 1789, and had thirteen months of four seven-day weeks. The months took one through from Moses (initial theocracy) through Homer, Aristotle, Archimedes, Cæsar, St Paul, Charlemagne, Dante, Gutenberg, Shakespeare, Descartes, Frederic II, and ended with Bichat (modern science). Every day has its patron, which include Buddha, Mahomet, Trajan, Alfred, Mozart and Montgolfier for the Sundays, and a host of lesser figures for weekdays, with alternates in leap years – Jesus is not commemorated. It is a fascinating document, in which due weight seems to be assigned to humanities. For many, however, Positivism seemed narrowly scientific, and its religion absurd, perhaps a cover for immorality, as it did to the schoolmaster-writer Edward James Mortimer Collins:[13]

> Life and the Universe show spontaneity
> Down with ridiculous notions of Deity!
> Churches and creeds are lost in the mists;
> Truth must be sought with the Positivists.
>
> Wise are their teachers beyond all comparison,
> Comte, Huxley, Tyndall, Mill, Morley, and Harrison
> Who will adventure to enter the lists,
> With such a squadron of Positivists?
>
> Social arrangements are awful miscarriages;
> Cause of all crime is our system of marriages;
> Poets with sonnets, and lovers with trysts,
> Kindle the ire of the Positivists.
>
> Husbands and wives should be all one community,
> Exquisite freedom with absolute unity;
> Wedding rings worse are than manacled wrists
> Such is the creed of the Positivists.

> There was an APE in the days that were earlier;
> Centuries passed, and his hair became curlier;
> Centuries more gave a thumb to his wrist, –
> Then he was a MAN – and a Positivist.
>
> If you are pious (mild form of insanity,)
> Bow down and worship the mass of humanity,
> Other religions are buried in mists;
> We're our own gods, say the Positivists.

Such suspicions of Positivism might reasonably attach to the lifestyle of John Chapman, who published Harriet Martineau's translation of Comte. He was also Marianne Evans' publisher as she went from translating German liberal theology into writing great novels under the name of George Eliot – and moved in with G.H. Lewes, a married man, meaning that she could not be received in respectable houses.

Positivism, with or without its Religion of Humanity, did make considerable noise among intellectuals in Britain as in France, especially in circles unhappy with orthodox religion. Associated with left-wing politics, its devotees were disappointed that few of the working class were attracted to the Religion of Humanity: they liked their Sundays free from both Christian and humanist preaching. Harriet Martineau, from an eminent Unitarian family, a long-time invalid who also famously wrote about her remarkable recovery, was a close friend of Charles Darwin's older brother Erasmus; and to connect Positivism with evolutionary speculation might seem plausible. Huxley though was indignant at being lumped, as he often was, among Positivists. This was partly because he lived an extremely respectable life, happily married and contemptuous of philandering. Believing that morality had no necessary connection with religion, he urged upon his fellow-agnostics the need to avoid the charge that their disbelief was a cloak for wrongdoing. Any public understanding of science, for which he believed that agnosticism was an important precondition, had to avoid that – as earlier it had had to eschew political revolution. Huxley was contemptuous of the Religion of Humanity, calling it Catholicism minus Christianity; and while he greatly admired David Hume's sceptical writings, he saw Comte's as misleading about science.

Inductivism, as Samuel Wilberforce emphasised, did not lead to evolutionary belief. It was necessary even for an agnostic (committed to doubt what could not be proved) to go beyond that, and to see Darwin's wonderful hypothesis, as Huxley saw it for many years, not as an improper leap in the dark, but as a light shining in the darkness. Following it, the scientist could bring into focus and relationship all sorts of things that otherwise seemed shadowy and unconnected. Huxley's own work, on classification and in particular the relationship of birds and reptiles, made sense in this new perspective, where taxonomy reflected ancestry. The imaginative leap that led him to perceive that iguanadon and other dinosaurs walked on their hind legs like ostriches, rather than crawling on their tummies like crocodiles, was not a feature of Positivist science. Evolutionary theory, so important in raising questions for research and in plausible explanations, did not measure up to the criteria for 'positive science' – it was not directly testable. Huxley was a great admirer of German universities, their research ethos and their laboratories for physiology, and had little time for the French Academy of Sciences (where evolution had a rough ride, and recognition for Darwin came late and grudgingly), or for Comtean philosophy; anything good in it had been in Bacon. Huxley's view of science as trained and organised common sense left room for hypotheses: after all, in ordinary life we use them when we aren't being positive about our knowledge.

Tyndall, the other man of science singled out by Collins, was a friend of Huxley's and fellow-member of the X-club, with its rather sinister reputation as an agnostic pressure-group. He had gone to Germany to do a PhD, and subsequently translated German scientific papers, notably Hermann Helmholtz's on conservation of energy.[14] Settled at the Royal Institution, colleague and then successor to the pious Faraday, he derived his strong feeling for the sublime from mountaineering. Davy and Faraday loved the Alps, being one of the first generation of intellectuals to derive pleasure from walking in the mountains, but Tyndall was a daredevil climber, making a number of first ascents of Alpine peaks. His ice-axe is preserved in the museum in Zermatt, and a side peak of the Matterhorn is named after him. To the alarm of Henrietta Huxley, he sometimes took his friends climbing, but he usually went only with local guides, and relished the opportunity sometimes to climb alone. He was a serious student of glaciers and their motions, but the major impulse behind his climbing was a devotion, leading to a kind of worship, to Nature: again, in the spirit of Davy and his rhapsodic poem.[15] This was nothing like the Religion of Humanity: Tyndall worshipped alone, on the mountain top. It was private, and he had no time for organised religion, or for supposed miracles. Science could, however, go well with personal spirituality, and required imagination. This vision he communicated to audiences at the Royal Institution and the BAAS, and in his writing about the Alps.

In tranquillity, Tyndall recollected his experience at the summit of the Weisshorn:[16]

> Over the peaks and through the valleys the sunbeams poured, unimpeded save by the mountains themselves, which in some cases drew their shadows in straight bars of darkness through the illuminated air. I had never before witnessed a scene which affected me like this.... An influence seemed to proceed from it direct to the soul; the delight and exultation experienced were not those of Reason or of Knowledge, but of BEING: – I was part of it, and it of me, and in the transcendent glory of Nature I entirely forgot myself as man. Suppose the sea waves exalted to nearly a thousand times their normal height, crest them with foam, and fancy yourself upon the most commanding crest, with the sunlight from a deep blue heaven illuminating such a scene, and you will have some idea of the form under which the Alps

present themselves from the summit of the Weisshorn. East, west, north and south, rose those 'billows of a granite sea' back to the distant heaven, which they hacked into an indented shore. I opened my note-book to make a few observations, but I soon relinquished the attempt. There was something incongruous, if not profane, in allowing the scientific faculty to interfere where silent worship was the 'reasonable service'.

Anyone who has climbed to a summit, especially in the early morning, will have exulted, though probably not quite in Tyndall's language. Heir, like his contemporaries, to both the Enlightenment and the Romantic movements (in his case, both English and German), he could draw upon the sublime sentiment in Wordsworth's writing and Caspar David Friedrich's painting,[17] using it to get the latest science across. Thus we find him doing quick calculations about expenditure of energy (where he was an expert[18]), drawing our attention to curious phenomena of refraction, and explaining just how glaciers move. They were in retreat, and a clue to understanding the Ice Age and the dark backward and abysm of time;[19] and were also a tourist attraction, as Thomas Cook invented the package holiday. More serious students could read all about it in Tyndall's somewhat more formal book (based upon lectures for children at the Royal Institution) on water, ice and steam,[20] with its splendid engraved frontispiece of a spiky mountain with a scarf of cloud streaming from it. A wider public would pick up much science, and a scientist's way of thinking, from reading about alpine exploits (as from other travel books). His theory of education was Emerson's, in which instruction was half the battle, and *provocation* the other; he did not teach a dry syllabus in a university (some who did looked snootily down upon his populism), and he had a wonderful way of getting through to diverse publics.

His Belfast Address had given enormous offence to some, bringing him great notoriety; but certainly achieved public understanding of some at least of the objectives, scope and methods of science, on its way to its powerful (he and Huxley might have hoped, dominant) place in our culture. His professorship at the Royal Institution and commitment to the BAAS made him a generalist in an age of increasing specialisation, and also propelled him into thinking about imagination. In 1870, at the British Association in Liverpool, he delivered a 'discourse' on the scientific use of the imagination. It was later published in his wide-ranging collection, *Fragments of Science*[21] – in the first edition the phrase 'for unscientific people' had been added to the title, in real or mock modesty, to indicate an audience. The discourse began with texts from Emerson and Goethe to set the tone, and picked up, from the Presidential Address (1859) to the Royal Society by the eminent physician Benjamin Brodie, an encomium upon the imagination – not rambling uncontrolled, but governed by experience and reflection, the noblest attribute of man, the source of poetic genius but also the instrument of discovery in science, without which Newton, Davy or Columbus would never have made their great discoveries. Tyndall planned his discourse in the Alps, reading poetry, logic and Goethe's work on colours, pondering on 'the self-inflicted hurts of genius, as it broke itself in vain against the philosophy of Newton'. Imagination not properly subjected to reason and experiment cannot be part of science – but 'Newton's passage from a falling apple to a falling moon was an act of the prepared imagination', and Tyndall's other heroes showed the same characteristic: 'the fact is, that without the exercise of this power, our knowledge of nature would be a mere tabulation of co-existences and successions.'

With ready use of an example from common life, raindrops falling into a pond, he took his audience through to the wave theory of light, and the luminiferous æther invoked so that there was something to wave, an equivalent of the pool of water; and went on to elucidate the blueness of the sky, and of Alpine lakes, and the redness of sunsets. The discourse is itself an example of imagination at work, shaping diverse experiences and explanations into a whole. He went on to discuss 'the doctrine of Relativity', meaning thereby the very different perceptions of the large and the small that the astronomer, the ordinary person, the microscopist and the atomic theorist would have: how we perceive things depends on our previous state. In Darwin's work,

> observation, imagination and reason combined have run back with wonderful sagacity and success over a certain length of the line of biological succession. Guided by analogy, in his 'Origin of Species' he placed at the root of life a primordial germ, from which he conceived the amazing variety of the organisms now upon the earth's surface might be deduced.

Such a hypothesis could not be final: the restless human mind wants to get back behind this 'germ'.

> In the dim twilight of conjecture the searcher welcomes every gleam, and seeks to augment his light by indirect instances. He studies the methods of nature in the ages and the worlds within his reach, in order to shape the course of speculation in antecedent ages and worlds. And though the certainty possessed by experimental enquiry is here shut out, we are not left entirely without guidance.

Philosophers and astronomers had come up with the nebular hypothesis long before there was good evidence for it: now it was part of the scientific big picture.

Tyndall moved towards the close of the discourse with an encomium upon matter, reminiscent of Priestley. If evolution (cosmic and terrestrial) were correct, then all our achievements (Plato, Shakespeare, Newton, Raphael) were potentially there in the fire of the Sun at the beginning of time. Matter is wonderful: it is not the brute, inanimate, cold opposite of spirit – rather, they are two sides of a coin, 'two opposite faces of the self-same mystery'. Goethe would have us look upon it as the living garment of God: and 'what God hath joined together, let not man put asunder'. He ended with a splendid peroration about evolutionary scientists and their extrapolations:

> Within the long range of physical enquiry, they have never discerned in nature the insertion of caprice. Throughout this range, the laws of physical and intellectual continuity have run side by side. Having thus determined the elements of their curve in a world of observation and experiment, they prolong that curve into an antecedent world, and accept as probable the unbroken sequence of development from the nebula to the present time. You never hear the really philosophical defenders of the doctrine of Uniformity speaking of *impossibilities* in nature. They never say, what they are constantly charged with saying, that it is impossible for the Builder of the universe to alter His work. Their business is not with the possible, but with the actual – not with a world that *might* be, but with a world that *is*. This they explore with a courage not unmixed with reverence, and according to methods which, like the quality of a tree, are tested by their fruits. They have but one desire – to know the truth. They have but one fear – to believe a lie. And if they know the strength of science, and rely upon it with unswerving trust, they also know the limits beyond which science ceases to be strong.

We notice the Biblical echoes here; and even that agnostics had been much interested in Henry Mansel's famous lectures on the limits of religious thought.[22] Tyndall's materialism (based on the wonders of matter, not the joys of shopping) caused a furore in those days when organised religion was such a powerful force in society, and when arguments tended to take religious form.

Tyndall popularised his own particular researches, but in lectures and writings like these, focusing upon imagination, also sought to increase public understanding of how scientists worked. This was at a time when there was interest in how far scientific method might be taught, instead of, or as well as actual science, in schools and elsewhere. As content got harder to popularise, to communicate to those who lacked mathematics or microscopes, so grasping something of scientists' world-view and big picture seemed a hopeful possibility. The only problem was that they did not all agree; but that made the project all the more fun for science-watchers. As Tyndall knew, and as Arthur Balfour was later to emphasise, science depended ultimately upon faith, notably in the uniformity of nature; and also upon matters of taste in world-making.

Three great unifying conceptions brought together the increasingly specialised sciences in the middle years of the nineteenth century; and could be imaginatively grasped by outsiders – even if to insiders it did not always seem that lay people got the right end of the stick. The first was the atomic theory whose story we have already met. Its roots were in Antiquity, to which the classically educated loved to revert; but the atoms of Democritus, Epicurus and Lucretius (revived in the days of Boyle and Newton) were very different from John Dalton's. Theirs were all composed of the same stuff, matter: his were distinct kinds, hydrogen, nitrogen, iron, zinc and so on, over thirty kinds and rapidly growing as more elements were discovered with new methods of analysis.

It seemed odd to many, including Davy, that very similar elements (like his sodium and potassium) should be irreducibly different, especially when the compound radical ammonium (NH_4) behaved very like them. Davy as President gave Dalton the Royal Society's Royal Medal in 1826, but in his speech made it clear that he admired the laws of chemical combination but not the speculative atomism that Dalton had invoked.[23] There could be no direct evidence for the existence of atoms; and there were analogical arguments in favour of seeing the chemical elements as composed of simpler particles. Dalton had proposed hypothetical arrangements, drawing atoms as little circles, and Wollaston arranged exact, flattened and elongated spheres into crystal forms; but these explained things only in principle, not in any testable detail. Most chemists regarded atoms as imaginary, fictions of occasional usefulness, part of the scaffolding rather than the structure of science.[24]

When Dalton's circle symbols were replaced by Berzelius' letters, seeing formulae as compressed recipes rather than structures became even easier. Problems arose when two different substances, ammonium cyanate and urea, turned out to have the same formula. If their atoms were differently arranged, that would explain their difference, and as further cases of this phenomenon of isomerism were found, atomic theory became more plausible. Nevertheless, it was no clear way of going from recipes to formulae. Dalton said that when one compound only was known, it must be AB: water was thus HO. Davy, Amadeo Avogadro, Andre-Marie Ampère and others, using volumes rather than weights, came up with H_2O. This had knock-on effects right through chemistry; its basic theory was a muddle, as different authors used different conventions. In 1860 an international conference called at Karlsruhe failed to reach agreement; but, in its aftermath, chemists went for the H_2O formula and all that followed from it. Soon the elements were being classified in what became known as the Periodic Table, conveying a great deal of information to those who already knew something.

It was still possible for the sophisticated to doubt the existence of atoms, though William Thomson had inferred their existence and size from the limited spreading of oil on puddles.[25] At the Chemical Society of London, there were major debates in 1867 and 1869, where prominent speakers, including a President, were among the doubters, urging a theory-free science, where thermodynamics might be invoked to account for isomerism. By then, August Hofmann had demonstrated his ball and wire molecular models, disdained by sceptics. Tyndall was invited to one of the debates; and he chided the chemists, then and in his discourse on imagination, with being over-scrupulous. Just as physicists accepted that light was waves, so should they accept that matter was atomic. Hermann Kolbe mocked the improved models, based upon Jacobus van' Hoff's insight of 1874, that the atoms grouped around a carbon atom formed a tetrahedron; the great physical chemist Wilhelm Ostwald held out against atoms into the twentieth century; but they were an embattled and even eccentric minority.

It seems to have been the demand for teaching in chemistry in the years after 1870 that led to the success of atomism: Dalton's imaginative leap went with general understanding in a way that abstract mathematical analyses never could.

The need to communicate drove theoretical understanding, however alarming this might seem to the cautiously logical – who, by the twentieth century, were being called 'positivists' (with a small p to show that they didn't go the whole way with Comte). The step taken, new realms of structural and synthetic chemistry, lay open to those whose powerful imaginations could envisage shapes and mechanisms.

Conservation of energy was another big idea, towards which many had been groping when Helmholtz, the great polymath,[26] gave in 1847 the famous lecture that Tyndall translated. He stated clearly and generally what others had seen more mistily, or as special cases, and perceived how this doctrine could unify sciences like mechanics, optics, electricity and magnetism – into what we call physics. Classical physics, taking over from chemistry as the fundamental science, made claims that stretched into chemistry, geology and biology. Its crucial premises were that matter and energy were both indestructible. The generation of an electric current from one of Volta's or Davy's cells, or from Faraday's dynamo, were examples of transformations of energy, which like different currencies could be exchanged. The big question became the exchange rates. Mechanical energy was measured in units of mass, length and time: centimetres, grams and seconds in the units becoming universal in science in Helmholtz's time. The task facing physicists was to express heat, magnetic, electrical, optical and even chemical energy in the same units. This involved precision equipment of the kind on show in those scientific exhibitions and museums flourishing in the late nineteenth century. From the 1850s a new coherence came into the physical sciences.

The progress of statistics, the study of uncertainty, bringing mathematics into the realm of chance, was a great feature of the nineteenth century;[27] and it became central in scientific explanation when, in 1859, James Clerk Maxwell announced his dynamical theory of gases.[28] This, the notion that they are composed of elastic particles bouncing off each other and off the walls, not only cohered importantly with chemical atomic theory, but also depended upon laws of large numbers. In a hot gas, the particles were on average moving faster than in a cold one; but there would be some at any given time moving very slowly, or even stationary. Boyle's and Charles' simple gas laws (concerning pressure, temperature and volume) worked not at the level of individual molecules, but only when there were millions of them – rather like death rates among humans. Maxwell helped public understanding, or maybe misunderstanding, by imagining a demon to illustrate how this theory impinged on ideas of time as well as determinism.

That energy is conserved is the first law of thermodynamics, the science of heat and work; and the second is that heat will only flow spontaneously from a hotter to a cooler body. This limits the efficiency of steam engines, means that you need to switch the fridge on, and indicates that the solar system will have a 'heat death' when everything has become equally tepid (or rather colder than that). It can be interpreted to imply that disorder, entropy, is always on the increase. It was thus the first law of physics that implied a direction and irreversibility to time: more entropy means later. If we have a divided container, one half of which contains a hot, and the other a cold gas, and we open a connecting flap, then in their random movements fast-moving molecules from the hotter side will go through to the colder, and slower-moving ones the other way, and the collisions will soon result in the whole mass being tepid. Order, and the capacity to do mechanical work represented by a temperature difference, has in time given way to undifferentiated disorder, equalising down. This can only be reversed by doing work, or so the second law requires.

Maxwell's demon was so small that he, or it, could see individual molecules. We cannot do that: his friend Thomson had ingeniously inferred their minute size, and we must work with big numbers.[29] The demon works the frictionless flap, permitting fast moving molecules to go one way, and slow-moving ones the other, as he sees them coming. Soon, he will have reversed the mixing process and restored the initial state, where one half was hot and the other cold: and demons do not require energy to keep going, or expend it in moving frictionless flaps. This imaginary situation, a 'thought experiment', indicates that this fundamental law of physics – seen by Charles Snow in his famous lecture on Two Cultures as the test of scientific literacy – is statistical. We do not have to imagine demons: it could be that spontaneously in our glass of warm beer fast-moving particles happened to collect one side, and slow-movers the other, so that it began to boil in one place and freeze in another. The fact that this is extremely unlikely is immaterial (after all, people do win national lotteries) in a discussion of causality: determinism seemed less strict than had been supposed, even in physics.

Herschel had described Darwin's natural selection as the law of higgledy-pigglety[30] (his being an orderly universe of 'verae causae', genuine causes); but in the very autumn in which the *Origin of Species* was published, 1859, Maxwell had announced his theory of gases at the BAAS. The spectroscope, and Maxwell's analysis, showed that molecules were all identical, and never wore away or broke in pieces. Animals and plants are different, and natural selection depended upon an informal statistical analysis: on average, and with large numbers, better-endowed individuals will survive in their environment and breed similar descendents. Maxwell, like his friend (and Helmholtz's) Thomson, was a devout man: Cambridge graduates like Herschel, they were socially superior to Huxley and Tyndall, and they disliked their cocky and perhaps windy agnostic 'Scientific Naturalism' with its rejection of anything supernatural as unknowable. Thomson applied thermodynamics to work out how long the Sun had been burning, and how long it might go on doing so.[31] Before energy conservation, nobody had worried overmuch about that; but now it was clear that the Sun could not yield all that heat and light from nothing. If it were made of the best coal, were shrinking on itself under gravity, and were being fuelled by meteorite impacts, then it might just have been burning for 100 million years, and might be expected to go on for about the same time. In fact, a half or even a quarter of those figures seemed more likely. Then the Earth was radiating heat into space, and it was possible to calculate how along ago it could have supported life: the figures came out about the same. Here then were two independent calculations that yielded convergent results – excellent evidence in science for believing that one is on the right lines.

These are huge numbers, and will see us out – though in his *Time Machine* of 1895 H.G. Wells made his readers shudder at the prospect of heat death, with a giant cool red Sun barely warming an Earth now only capable once again of supporting monstrous invertebrate creatures and slimy weeds in its thin air.[32] But while further back than the 4004 BC creation-date of Biblical literalists, Thomson's figures as he knew were insufficient for the undirected geological processes envisaged by Lyell, Darwin and Huxley.[33] The leading physicist and the leading biologist in Britain found themselves publicly at odds. This is the kind of thing that always helps in the public understanding of science – the assumptions, pecking-orders and world-views are brought out, and the issues involved have, willy-nilly, to be made intelligible to those outside narrowly specialised and professional groups. Huxley's famous encounter with Wilberforce seems to have been played down by those responsible for running the BAAS;[34] they wanted science to be perceived as exemplary and rational, not to be the subject of witty or bad-tempered debate (like politics). Similarly, Huxley's clashes with Richard Owen were deplored as prize-fighting by a commentator:[35]

> It matters little to us, whether or not the strife continues; and as far as the public are concerned, they take it as a matter of course that Professor Owen will be attacked whenever Professor Huxley speaks or writes; or they crowd into the lecture hall with the same feelings as they would go to witness a prize fight; all we can say is, that it imparts to the non-scientific world a false estimate of the spirit which exists amongst scientific men, a very false estimate indeed, and what chiefly concerns us as reviewers is that it does great permanent injury and reduces the intrinsic value of an author's works, for it is difficult to accredit a writer with strict impartiality, who cannot exercise a little control over his feelings.

We cannot but imagine that it would have made excellent television. With Thomson, whose mathematics was way beyond Huxley, the quarrel was conducted more politely.

Evolution was the other big idea which unified a vast range of sciences, as the chapter-headings in *Vestiges* and in the *Origin of Species* indicate. The former starts with astronomy, works through the eras of the Earth's history, and ends with psychology: the latter takes us from stockbreeding through instincts, hybridism, geology, geography and taxonomy, with a whole chapter devoted to the theory's difficulties. Specialised sciences had become less attractive to outsiders, more akin to Thomas Kuhn's 'normal science', than they had been about 1800; but this eye-opening and imaginative account of how everything was connected was irresistible to all sorts of general publics. Gorilla cartoons, anxious sermons, the placing of different human groups in a scale of races from the primitive to the modern, were popular at all levels. That does not mean that everybody was convinced: far from it. It was just that evolutionary ideas became so fashionable as to be almost unavoidable; a focus for belief or ridicule, but part of the Zeitgeist.[36] Empire-builders foresaw, sometimes complacently, the inevitable fading-away of peoples less well-equipped for the struggle than theirs; William Crookes explained, as the result of inorganic evolution, the baffling number and relationships of the chemical elements; Norman Lockyer saw stars evolving and passing through various stages.

Darwin's open-ended struggle for existence, that competition red in tooth and claw for the best niches and the chance to leave descendents, was less attractive to most than progressive evolutionary views, in line with Victorian confidence that things were getting better. Darwin's barnacles had gone down in the world, and thereby increased and multiplied, flourishing greatly. There was some feeling among the glumly imaginative that something like that was happening among humans. Degeneration seemed, when one looked, to be everywhere. There were throwbacks, backward individuals of all kinds; and degenerates, who could be expected to swell the criminal underclass. Building upon the phrenology which had seemed discredited, the alarmed could see everywhere real or potential villains and offenders. Moreover, these people were breeding faster than middle-class intellectuals, who were beginning to limit their families. The last decade of the nineteenth century was not only the Naughty Nineties, but also a time for fin-de-siècle gloom, a stiff absinthe, and maybe investing in a swordstick or a dagger cane. The answer, it seemed to Francis Galton and others, was eugenics; but, in the short term, there was as much scope for doom and gloom as for celebration of the new century.

Evolution was, as one would expect with something developing from natural history, highly accessible to a general public who were ignorant of, and not very interested in, more detailed studies like Darwin's of barnacles – though anyone could appreciate his book on earthworms. Thermodynamics never achieved that kind of resonance. But steam engines lay behind that science, with the question of how much work could be expected from a certain quantity of coal as Smeaton, then Watt, and then the Cornish engineers Trevethick and Hornblower improved enormously upon Thomas Newcomen's prototype. Stationary, shipboard and locomotive steam engines, and then great turbines, followed; and great machines stirred the imagination and roused the pride of Victorians. Samuel Smiles might put success in engineering chiefly down to the stern virtues of self-help, character, thrift and duty; but technology like science is the fruit of imagination, controlled by the way things are. It is constrained by costs, and by time limits. Now there are those who seek to separate science and technology from each other, but in the nineteenth century this conception of 'pure science' only began very late. Science might indeed be intellectually exciting, but a major part of its attraction then – and indeed for most of us still – was that it was useful. It would, as Bacon foresaw, ease labour and cure diseases.

By the 1870s, with the formal teaching of science on a large scale, seeing technology as 'applied science' began to look much more plausible than it had before. New industries were based upon electricity and chemistry: steam engines had led to thermodynamics, but Faraday's experiments with little coils of wire and magnets in the laboratory led to electric motors. Synthetic dyes were the fruit of the Royal College of Chemistry and its research programme. In a world snobbish about where money came from, the scientist was a professional, while the technologist was making and selling things. But William Armstrong was made a peer for his work on breech-loading guns and warships, and William Thomson for his work on the Atlantic telegraph especially: technology was by no means despised. And whereas in the 1840s the pioneering engineering course in the University of Durham (though half a century behind the École Polytechnique) failed because employers were not interested in paper qualifications, by the turn of the century technical education in Britain was flourishing greatly.[37]

The century had begun with cautious Baconian assumptions about the scope and method of science as careful inductive generalisation, playing for safety in a dangerously revolutionary age. It ended with big imaginative syntheses, and a widespread enthusiasm for hypotheses, deductions and testing as the best scientific procedure. The centre of science in 1800 had been Paris, capital of a formidable military power but essentially pre-industrial. By 1900, science could not be divorced from technical progress, which led to the sophisticated apparatus that made 'pure', blue-skies science possible. There was now a big scientific community – it read journals like *Nature* – but there was also room for others much less formal. And it is to *Science Gossip*, its predecessors and its contemporaries, that we now turn.

10 Science gossip

The nineteenth century was the great age of the periodical. Up to 1800, printing had not changed very much since the days of Gutenberg and Caxton: type was set by hand, and the press worked by human muscles, producing small editions for those who could read. There were not very many who could read, but by 1800 elementary education was, even in backward England, increasing the number of readers; while gas lighting, and improved oil lamps on Argand's principle, with cylindrical wicks and a good air supply so that they did not smoke, meant that study after dark became much easier by the 1820s. This was the era of the evening class and the Mechanics' Institute. Knowledge is power, and new classes wanted it. Extensive reading replaced the intensive kind, where readers had returned over and over again to a few texts like the Bible and *Pilgrim's Progress*. There was a big market in prospect for printed matter, a hunger for enlightenment that alarmed the well-off because revolutionary, 'Jacobinical', books and pamphlets threatened the social order. But the invention of the steam press made practicable very long print runs that would have taken weeks on hand presses, and thus opened a door to economies of scale and cheap periodical literature. The publication in parts which had been a feature of fine natural histories became the norm for novels like Dickens' – read breathlessly in instalments, each one (like the *Arabian Nights*) ending at an exciting moment; and then published after some revisions as a respectable book.

Associated with steam presses was stereotyping, in which a cast was taken of the type after setting up – the type was then distributed for re-use, and the stereotype could be used for a long run, or over and over in reprints; and the cliché, a cast from a wood-engraving, which again could then be used for large-scale publication. Curiously, these terms from fast-changing new technology became metaphors for fixity. Now, cheap publications could thus satisfy Alice in Wonderland, having pictures as well as conversations to draw the reader in. Restrictions imposed especially in the post-war clamp-down after the Battle of Waterloo in 1815 (licensing, censorship and taxes, 'stamp duty') and libel laws threatened the editor and publisher of cheap periodicals, and some of them did find themselves in gaol; but these 'taxes on knowledge' were lifted in Britain by the middle of the century, and were absent in the USA. All this meant that informal, religious or radical or conservative, populist publications became extremely widespread and important – and that science, as an increasingly prominent part of culture and engine of technology, found a place within them.

A great scholarly enterprise, the Scientific Periodicals Project, has just resulted in the publication of three books full of learned essays dealing with just our theme – popular science in journals directed to various readerships.[1] Even this band of scholars can only sample the enormous variety and quantity of material, most of it not helpfully indexed, so that hefty volumes have to be painstakingly gone through. As they stress, there are great runs of journals still awaiting attention – they have looked for representative titles, and different decades. It is a little curious to see such scholarly industry put into publications which were designedly ephemeral, intended as journalism responding to current needs and interests, and written to deadlines. But such lively and relatively artless writing, which depended upon getting and holding on to a paying readership, can illuminate an intellectual climate in ways that the austere publications of learned societies supplied to members and academic libraries cannot. Much science, more then than now perhaps, is published with posterity in mind. Journalists have different priorities, and so do popularisers. Though masterpieces by Dickens and other great novelists are found in popular journals, nobody expected to place or find original works of science there. What we do find, however, from time to time is that major scientists did popularise their own work, and that of contemporaries, in these sometimes-surprising publications. This might bring them into entertaining or serious controversy. We also get some idea, through journalists, of popular reactions to science, and of what caught the public imagination.

Looking at periodicals published by religious bodies, at the comic journal *Punch*, at journals aimed at women, at children, or at intellectuals, the contributors to the project make some valuable generalisations. They come out firmly against the notion that the public understanding of science is simply to be understood as some kind of trickling down of information about the latest discoveries from Fellows of the Royal Society to admiring people on the Clapham omnibus. The various publics all have their interests, and those who write for periodicals must catch and hold their readers. Only some bits of science will help them, and very likely not the authorised bits. Rows are often helpful to journalists, and so are obvious effects on people's everyday lives. The journals were indeed conduits, but they were also interactive, with editors trying to relate to, enthuse and perhaps mobilise their readership; and they possessed different degrees of authority. The project showed the great diversity of the field. Nineteenth-century editors and contributors responded to science enthusiastically, respectfully, mockingly, timidly or fearfully; and prominent from time to time in general journals were debates, controversies, copyright disputes, plagiarism and misunderstanding, as well as broader or narrower perspectives on science than those of the experts. With popular works like *Vestiges*, the range of reviews and comments helps us to get a proper view of how it was received and why it was so sensational.

Some journals had and have long histories, celebrating centenaries;[2] others, as the project's authors bring out, had short but important lives. Failure may be more instructive than success, at the time and for the historian; also, something absolutely adapted for a particular decade and very influential then, may wither and decay after its flowering without the stigma of failure. Conversely, stately vessels that in the past have weathered storms and sailed in strange seas of thought may drift on for many years into the present hopelessly becalmed in doldrums. The project's authors bring out themes like nationalism and imperialism; intellectual property and anonymity; contributions by women; spiritualism; and costs, prices and distribution, indicating how, with railways and telegraphs, London became ever more important as a centre of intellectual, cultural and publishing activity. We learn much about the complexities of Anglo-American publishing before international copyright was established, about co-publishing, and the circulation of British publications across the Atlantic.

The huge importance of religion, religious publications and the necessity of maintaining a respectable tone is made evident throughout. *Punch*, with its cheerful debunking style, nevertheless remained respectable; the 'new journalism' of William T. Stead, suspicious and interrogatory, with its proactive and sensational character and cheeky distrust for authority, did not. Among the agnostics, Huxley was respectable; Tyndall, with the Belfast Address of 1874, teetered on the brink; but William Kingdom Clifford, a zealot happy to discuss taboo topics like prostitution and divorce, who mocked religion and relished the 'pagan' poetry of Algernon Charles Swinburne, went too far, and his papers had to be carefully edited, or filleted, when they were re-published in book form after his early death.

Clearly, support for science was essential from fickle publics who were, one way or another, expected to pay for it. Popular periodicals of one kind or another made science more familiar, and its leading practitioners public figures, perhaps sages. In his journal, *The Nineteenth Century*, the architect turned editor Sir James Knowles set his literary lions, including Huxley, W.E. Gladstone and other prominent figures in science and public life, to maul each other for the delight of his readers. Perhaps really it was more of a fun-fight than a battle, but the issues might be serious.[3] Gladstone had hoped that the narrative in *Genesis* 1 might in a general way be compatible with current geology, but Huxley put him right with witty remarks about the remarkable elasticity of the Hebrew language; and when Huxley died, he was struggling with a review of Arthur Balfour's book presenting science as based ultimately upon faith.[4]

At the beginning of the nineteenth century, the *Edinburgh Review* had transformed the reviewing of books, including scientific ones. It became the example followed by other publications, such as the *Quarterly*, the *Westminster* and the *North British*. All these were dedicated to one high culture, discursively reviewing in lengthy essays works of travel, politics, literature, religion and science. They also adhered to a culture of anonymity: contributors, sent copies of the books for review, submitted their essays which might be heavily amended by the editor.[5] Sheltered by anonymity, friends (even sometimes the author himself) might puff, and enemies might damn, a book. Often it was an open secret who had written a particular review, as we can see from Darwin's correspondence after the *Origin of Species* was published, but a reviewer might be misidentified, and thus someone wrongly believed to be a friend or an enemy – and the editor might have changed the writer's tone. There were some surprising choices, as when John Murray, publisher of the *Origin*, invited Samuel Wilberforce (hardly expected to be sympathetic) to read the book for the *Quarterly Review* and duly got a hostile essay.

Editors did not mess about with the text of established writers like Macaulay, or indeed Wilberforce; and reviewers might in due course publish their essays from the *Reviews* in a book with their name on the title-page.[6] Anonymity allowed authors to equivocate when challenged, and bred mystery, but the *Reviews* were the way authoritative literary, philosophical or historical essays could be published (there being no learned journals in humanities or social sciences), they paid well, and could bring reputation.[7] It seems to have been the French who believed in signed reviews; and during the second half of the nineteenth century, this gradually became editorial policy in more and more journals.[8] Malice or innuendo was still, and is still, possible, but the probability of such bad faith was reduced.

Scientific papers have, from the beginning, generally been signed, though the experimental reports of the Accademia del Cimento in the 1650s were not: they were presented as collective, vouched for by the membership.[9] This Baconian ideal was soon abandoned: reviews, whether of books or of progress in some domain, often remained anonymous; but research papers to be credible had to be signed, and in principle, at least in the more august journals, verified by an anonymous referee who would as far as possible repeat experiments, check calculations and look up citations. Such 'peer review' caused trouble between Newton and Robert Hooke, to whom Newton's first paper on light and colour was sent. It is still an awkward business but, like democracy, it is the best way there is.

Newton's paper was in due course published by the Royal Society, and was therefore aimed at a small international public. But the Society's journal, *Philosophical Transactions*, was, in the seventeenth century, informal. It carried book reviews, and many of the papers were essentially letters to the editor: indeed, the scientific journal emerged from the correspondence network. By the end of the eighteenth century, it had increased in size to handsome quarto format, on excellent paper, with beautifully executed engravings. It was expensive, and radiated authority. The journal of the Paris Academy of Sciences was notoriously slow in publication, giving rise to all sorts of problems in settling priority disputes: *Philosophical Transactions* was better. But the reviews had been dropped, there were no reports of the Society's meetings or of papers read there but not published. Those papers that were published, and which generally read well, represent on the whole tidied and completed pieces of research rather than work in progress. The tone is gentlemanly, and controversy was minimised, though little triumphs (especially over foreigners) were allowed. Those who were not Fellows of the Society had to submit their work through a Fellow; and papers were usually read very formally by the Secretary, while Banks presided in court dress sitting behind the mace.

Philosophical Transactions still continues in its august way, but after publishing at the beginning of the nineteenth century a large and handsome edition (edited and abridged with papers in Latin or French translated) of the complete run, the Society decided to continue the process, and in 1832 published abstracts of papers from 1800–14.[10] As they caught up, other items of interest to Fellows found their way in, and by volume seven (1854–5) the journal's transformation was recognised in its new title, the Society's *Proceedings*. By then, this relatively informal octavo publication included less-finished research papers, speeches by the President, abstracts of papers read, short papers, news of the Society, letters, notes and obituaries.[11] This journal also continues to this day, but its history has been typical of what happens to scientific periodicals: starting informal, they gradually become more and more austere. Nowadays, the *Notes and Records* and the *Biographical Memoirs* augment the highly respectable and objective *Proceedings*. Many scientific societies also now publish newsletters, perhaps in glossy magazine format, to maintain informal communication, separating serious science (represented by peer-reviewed papers in unreadable prose) from the subjective, the personal and the chatty. Although such newsletters, which include publications like *Chemistry in Britain* and *Chemistry and Industry*, come from learned societies and aim to get across the latest science in accessible form, this difficult task is not highly regarded professionally. In Research Assessment Exercises, or at promotion committees, essays and reviews like theirs carry negligible weight. This is a change from the heyday of Herschel, Faraday, Huxley and Maxwell, whose popular writing brought them esteem.

At the beginning of the nineteenth century, the separate parts of *Philosophical Transactions* came out (like many books) in blue sugar-paper wrappers, ready to be bound up at the end of the year into a volume. Then someone had the idea of printing useful information, about availability of back-numbers for instance, on the wrappers, which were changed to a lighter, buff colour so that the print was readable. Sometimes important notes from authors might be there.[12] By the twentieth century, sober-coloured issues of current periodicals were a feature of shelves or tables in libraries, conversation-rooms and at conferences; and the amount of information on the wrapper steadily increased. The names and addresses of the editor and his board of assistants, and of the society or commercial publisher that brings it out, and the house-style for references and bibliography, generally now accompany the list of back numbers and prices on the covers – usually thrown out when sets are bound. When the historical journal *Annals of Science* was launched in 1936, its editors had the bright idea of making the cover brilliant orange so that it would stand out. It still does, though their example has been followed by others seeking to look dashing rather than sober, and library tables are nowadays more interesting than they were.[13]

By the early nineteenth century, the Royal Society allowed authors of papers in *Philosophical Transactions* to order offprints, and these became an important means of circulating information (as reading everything published became impossible) among scientists.[14] Certainly by 1807, offprints were available, loosely stitched in a plain blue sugar-paper wrapper, with the pages renumbered starting at 1. On the back of the title page there appears the injunction:

> Gentlemen who are indulged with separate Copies of their Communications, are requested to use their endeavour to prevent them from being reprinted, till one month after the publication of that Part of the Philosophical Transactions in which they are inserted. *By Order of the President and Council*

The message is duly followed by the name of the current Secretary of the Society, W.H. Wollaston, then W.T. Brande. By 1828 that has disappeared, and the pagination is the same as that of the paper in *Philosophical Transactions,* so it could more readily be cited. By the later part of the century, offprints like other printed matter could be posted for one halfpenny.

The fear about papers being reprinted had to do with less-exalted journals, run commercially, which from the end of the eighteenth century were providing a service more closely connected with public understanding of science. Scientists are and were aware of the pecking-order among journals – and only the respectable ones were cited in the massive (and international) *Royal Society Catalogue of Scientific Papers*, about which we are told in a little flyer of 1872 that five volumes were available to the public, at 28/- (shillings) each in morocco, or 20/- in cloth; but to Fellows of the Society at 18/6 or 13/4.[15] But we are interested in public understanding of science, and it is not to these fountain-heads of supposedly established knowledge that we shall turn but to the humbler vehicles concerned with its dissemination. They aimed to get science, as fast as possible, among those whose enthusiasm was more important than the social or educational position, lack of leisure, or remoteness that cut them off from major and affluent scientific societies. Their editors and publishers also sought a relationship with their readers, building up loyal subscribers into an 'invisible college' like the correspondence network to which Robert Boyle had belonged.

In the German states, a chemical community had been built up by the late eighteenth century by Lorenz Crell through his journal, *Chemische Annalen*.[16] This brought together a very diverse readership, involved in alchemy, pharmacy, medicine, or just curious about nature; and made chemistry an exciting field to be involved with. Crell's plan had been for original, comprehensible papers, translations, reviews and perhaps research proposals; and for selling not only through booksellers, but also by direct subscription. Lavoisier and his associates also started a journal, *Annales de chimie*, to propagate his oxygen theory against the hypothesis of Becher, Stahl and the Germans that phlogiston was the cause of burning; but theirs was edited by Academicians, intended for an expert readership,[17] and soon became the place to publish chemical discoveries.

Aware of these foreign examples, in 1797 William Nicholson, who had written and translated textbooks especially of chemistry, began a *Journal of Natural Philosophy, Chemistry, and the Arts*, which came out as a small quarto. With book reviews, reports of meetings, reprints or abstracts of papers in more august publications, and letters, it also offered the possibility of rapid publication of work in progress, curious experiments or exciting discoveries that were not yet sufficiently organised and understood to be reported to the Royal Society. *Nicholson's Journal*, as it was always called, secured priority; and its readers saw the minds of great men, like the young Davy, gradually gaining ground upon the dark – in a partly-collective enterprise as readers chipped in with their ideas and experiments.

Especially in new and unexpected fields, like electrochemistry opened up by Volta's electric cell or battery of 1799, readers could readily join in; after all, the apparatus was easy to make at home. The editor might also comment, and draw attention to new developments in any branch of science. In 1798 another journal, *The Philosophical Magazine*, was launched by Alexander Tilloch, a Sandemanian sympathiser, and the reviver with Andrew Foulis of Glasgow of the technique of stereotyping – they made it much more practicable, but made little money thereby. This journal was a direct competitor with Nicholson's, which adopted the same octavo format in 1802, but which was swallowed up by *The Philosophical Magazine* in 1813. In 1822, Richard Taylor joined Tilloch in editing the journal, which continues to this day, published by Taylor and Francis.[18] As scientific journals tend to do, it became more formal. By 1840, it had a team of three editors and two assistants, appeared monthly, and contained a diverse collection of papers, some translated and some reprinted, with accounts of meetings of scientific societies, book reviews, abstracts of papers, weather reports and descriptions of industrial processes. Readers would get a good conspectus of the science of the day, with an international flavour, and find contributions from very eminent authors like the geologist Charles Lyell, the mathematician James Sylvester, and the chemist Robert Kane. They might puzzle with Peter Mark Roget over the problem of moving the Knight over every square of the chess board, without going twice over any one. The papers were on the whole readable, and were illustrated with wood-engravings and copper-plates; but in some, familiarity with the differential and integral calculus was taken for granted, and others used formidable terminology. Most readers however would have lost the feeling that they were all part of a big team, and that all of them alike had a role to play in the great enterprise of science. By 1897, *The Philosophical Magazine* had become a physics journal, and it was there that J.J. Thomson published his classic paper on the electron.[19]

In 1813 another Scot, Thomas Thomson, began another journal, *Annals of Philosophy*, again appearing monthly and offering much the same advantages as *Nicholson's Journal* and *The Philosophical Magazine* had. He was a keen chemist, the first to realise the importance of John Dalton's atomic theory, which he promoted in his classic textbook. Elected Regius Professor of Chemistry at Glasgow, he introduced laboratory teaching for his students in the medical school. He became notorious for preferring their analyses, giving whole-number figures for atomic weights in accordance with William Prout's hypothesis (published in his journal) that the elements were polymers of hydrogen – something

J.J. Thomson commended in 1897, but which looked slapdash in the 1820s when compared with J.J. Berzelius' scrupulously un-rounded data.[20]

In his journal, he included biographies of prominent men of science – British and foreign. He also sought to include 'scientific intelligence', a survey of the previous year published each January keeping readers up-to-date, and new patents; and reported on the Geological Society as well as other more general learned bodies. Once again, readers would have been well informed about science, but the journal was hardly an open forum as Nicholson's had been in its early days. The physical sciences were becoming more technical, and theory in chemistry was more settled: it was harder to jump in to the scientific pool. By 1818, Thomson had acquired two assistants, and there was an impressive proportion of papers from abroad; but by then his duties in Glasgow were taking more time. After beginning a new series in 1821, the journal was in 1826 duly swallowed up by the omnivorous *Philosophical Magazine*, which also absorbed the journal that David Brewster (biographer of Newton, and inventor of the Kaleidoscope and other devices) had published in Edinburgh. Inexorably, it seems, periodicals that had set out to be popular and accessible became (in order to survive) formal and respectable.

Ten years on, Thomson assisted his nephew Robert, a doctor later distinguished in the field of public health and a Fellow of the Royal Society, in editing *Records of General Science* beginning in 1835. The objectives were very much the same as those of *Annals of Philosophy* had been, with biographies, scientific intelligence, book reviews, translations and original papers. What is particularly striking is the long article by Thomas Thomson, broken up as was usual in all these journals into monthly parts, on calico printing. This was a field in which science was augmenting practical experience: well before the introduction of synthetic dyes, chemists had been busy finding and improving natural dyes, and making them adhere to different fibres.[21] The journal has samples of dyed fabrics, about three centimetres square, pasted to the pages of the article as vivid examples of what was being discussed. Some of the colours have bled onto the facing page (especially the logwoods), but most are still fresh and give the lie to the notion that until synthetic dyes came along our ancestors' lives were drab. The emphasis in this journal was on the practical, and on aspects of chemistry, which was by this time becoming one of the most useful of the sciences. The journal, like some others, used two different sizes of type, a smaller font for the reports than for the main articles. It would seem to have been aimed particularly at bringing an understanding of science and its value to those involved in manufactures, who might be uneasy about book-learning and theory. It sought especially to bring foreign work before the British public.

An earlier, and also rather short-lived, journal focused upon useful science was the *Register of Arts and Sciences*, which began in 1824 and appeared every two weeks at the low price of three pence (weekly publications were subject to stamp-tax[22]). It was thus addressed to a less well-off readership: artisans and mechanics, rather than middle-class. The type is small, and there are no copperplates, but each issue had a full-page wood-cut as its front page, and others inside – 'upwards of one hundred' in each of the first two volumes, including an ingenious fold-out plate with a long, thin illustration. Inventions, discoveries and processes were a particular interest: readers could find out about Davy's work on an azure pigment (seeking to replicate a Pompeian colour), an oxy-hydrogen blowpipe, a monorail, a telegraph system, a cooker for ships and a boiler that consumed its smoke. They were warned about the dangers of keeping milk in lead vessels, and the evils of the treadmill were denounced; it should be:[23]

> Considered less in the light of a mechanical engine than in that of an instrument of torture, befitted better to the black purposes of despotic rule than to the penal visitations of a noble, free and civilized land; and, we frankly confess, that in noticing the thing, it is less intended as the object of mechanical study, than for the purpose of asserting, as Englishmen should do, our abhorrence of its use, as the inflicting hand of that justice, which, it is the proud boast of England, is so tempered with wisdom, propriety, and mercy ... *to the weaker sex in particular, a most scandalous and grievous oppression.*

Other devices described, in more welcoming tone, included Count Rumford's method of keeping coffee fresh, and Davy's electrochemical 'protectors' for the copper bottoms of warships. Medicine featured, with the horrific account of the dissection of the corpse of an executed felon which we met earlier; and also with Prout's discovery of free muriatic (hydrochloric) acid in the stomach as an agent of digestion. The volumes were efficiently indexed; and were much more open than *The Philosophical Magazine* to letters, signed or anonymous (from PAM or IGNORAMUS, for example). Clearly the editors, who are not named, sought to build up that relationship with readers so vital in popular publication. What is striking, comparing these publications with modern glossy newsletters, is the absence of advertisements: they had to pay their way through sales, covering the 'great cost and much difficulty' experienced in getting original articles, as well as the editors' bread and butter.

Aiming at a wealthier readership, primarily no doubt those who went to the lectures, was the *Quarterly Journal of Science, Literature, and the Arts* edited at the Royal Institution during these years. This was published by John Murray (whose office was in the same street) in octavo, well-illustrated (occasionally in colour) and including readable articles, mostly much longer than in the *Register* and much less cramped. Fields as diverse as astronomy, geography, geology and mineralogy, medicine, inventions, and descriptions of apparatus and instruments were covered in its original articles. There was a good deal of 'miscellaneous intelligence' (for which the Royal Institution was well-placed, as we know from its library catalogue[24]), translations and abstracts, with the usual reviews, notes and letters: there are also prospectuses for Brande's lectures on chemistry, with its associated textbook. By 1825, Faraday's name as lecturer was associated with Brande's. We find also Faraday's picture of a portable laboratory, accompanying an article on the analysis of mineral waters, and estimates of the height of the Himalaya mountains, establishing that Dhawalagiri was some 28,000 feet high – thus exceeding Chimborazo in the Andes, which when Humboldt climbed high on it was supposed to be the world's highest. There is an illustrated paper showing the advantages of curvilinear rather than the traditional square sterns for warships, as better to defend; and a long review of a book on ancient and modern wines. A learned parson, the Reverend G. Swayne, contributed a paper on the manufacture of British opium:[25]

> It should seem that the eyes of the British public are at length beginning to open to the prospect of those advantages which would be likely to accrue to the community, from the introduction of an article of commerce, so much wanted at home, to supersede the abominably adulterated drug with which the guardians of our health are supplied from the Levant, &c., under that name; and so much in demand, that the last advices from India inform us, that whilst trade in almost all other articles was in an unusually depressed state, the price of opium had risen from 20 to 25 rupees per chest.

He described a simple apparatus for collecting the poppy-juice, with which on one day in 1818 a single individual, over seventy years old, had collected five and a half ounces of the precious fluid – crucial at that time for pain-relief, and its addictive quality not fully realised despite the misadventures of Coleridge and others.[26]

By subscribing to this journal, the already well-educated would be able to keep up with science; and it had more informal exchanges between the editor and his public than *The Philosophical Magazine* at that date, with queries being answered in a section addressed at the beginning to 'our readers and correspondents'. The Royal Society's *Philosophical Transactions*, like other grand publications, would not print work which had appeared anywhere else (which meant that readers might well never hear of it): the *Quarterly Journal* was not like that, and reprinted a biography of William Henry first published by the Manchester Lit. and Phil. Clearly, quarterly appearance made it less suitable than the monthly *Philosophical Magazine* or *Annals of Philosophy* for those anxious to establish priority, and we see less science advancing dialectically through its pages than in the early days of *Nicholson's Journal* when electrochemistry was young. The various exchanges with readers, and the scientific intelligence, do not perhaps constitute 'science gossip'; and the days of a journal with that title were still far off. Science in these publications was widely interpreted, but taken seriously, and there is little chit-chat.

By the 1820s, specialisation was coming, even into more popular scientific journals. *The Chemist*[27] was aimed at artisans, and was scathing about Davy's social pretensions as President of the Royal Society; but paid its contributors and had a very short life indeed. More successful was *The Mechanics Magazine*, which ran for half a century[28] with largely practical articles; and *The Magazine of Natural History*, edited by John Claudius Loudon and published by Longman. The interest in, and market for, natural history and gardening was booming right through the nineteenth century (and beyond, into our own day); and the amateur was and is still able to participate much more fully than in laboratory sciences. Loudon set out his intentions in a preface to the second volume:[29]

> The grand object with which we set out, that of promoting a taste for natural history among general readers, and especially among young persons, has been steadily kept in view.... Throughout the work the subjects are treated with sufficient technicalities for the purposes of scientific accuracy; but at the same time so as rather to invite the stranger to these studies, than to deter him from them. As the taste of our young readers becomes more refined and critical, it will demand articles more rigidly technical and profound, and we shall not then be wanting in affording a supply. In the mean time, our correspondents may regard themselves as cooperating in a Magazine of their own, for the improvement of one another, as well as for the benefit of the public. To those who are impressed with the importance of natural History, as a means of educating the feelings and the heart, it must be satisfactory to know, that this science is spreading among all classes, and that Natural History Societies, Museums, and Libraries (we wish we could add Public Botanic Gardens) are formed, or are forming, in many of our provincial towns.
>
> The great use of Natural History and Comparative Anatomy is to humanise and soften the heart. If boys were acquainted with the wonderful structure of insects, and other animals low in the scale, they would not be found sticking pins into flies, or tormenting cats; nor, when men, would they treat those noble domestic animals, the horse and the ox, with cruelty. The girl who has learned to derive enjoyment from observing the operations and watching the metamorphoses of insects ... will learn also the elevation of her own nature. As she grows up to womanhood, she will feel more intensely the delicacy and dignity of the female character, and resist with force the temptations which always beset innocence, amiability, and inexperience.... The mind rationally occupied with the study of nature will no longer seek relief from *ennui* in bad novels; and the same superior taste for information, and the same admiration of the wisdom of Nature, as displayed in her works, will lead to a more select choice of companions, male as well as female.

These advantages demanded scientific study, just as manufacturers required machinery: and that the journal would seek to supply. Clearly, Loudon had in mind a middle-class readership, though there were groups of working men who botanised regularly. No doubt, the urge for 'improvement' as well as curiosity drove all of them in a world of booming industrial towns, and nostalgia for the lost world of Merrie England. The journal carried on its title page an engraving symbolic of its contents: Linnaeus' bewigged effigy on a medal with rays streaming from it, surrounded by a tortoise, two eagles, a giraffe, a lion struggling with a snake, and a fish; there are also fossils, shells, birds and a bat filling up the space, with vegetation suggested in the background. In the third and later volumes, this design was replaced by what looks like a large lifebelt with the words 'naturalists and lovers of nature' on it, surmounted by a bust of Linnaeus and with 'VOL. III. 1830' in its plain white middle, and two previous illustrated volumes open at the bottom, against a landscape with cliffs and water, trees at the top, and animals and birds round the sides.

The inside was lively. There were queries and answers, reports from various counties about wildlife and meetings, and a caustic anonymous contribution on the advantages and disadvantages of natural-history periodicals (which tells us a lot about how publishing worked).[30] In addition, there were papers on the luminosity of the sea and a caustic note about an obituary of Davy, like a portrait without warts and all. Dr Paris:

> Seems to wish, by implication, to puff off Sir Humphry Davy's piety; his *real* sentiments on religious subjects were well-known to his intimate friends. The inference which Dr Paris would wish his readers to draw respecting the domestic life of the philosopher is, that Sir Humphry Davy and his lady lived in a state of the highest connubial felicity. On this subject he had done much better to have maintained silence.

This is just what the historian, or the contemporary after science gossip, wants to know. Amid book-reviews, anniversary addresses, musical scores of bird songs, and weather reports, there were sometimes furious controversies. Thus the testy illustrator William Swainson was provoked by attacks on French naturalists who had presumed to name the Malay tapir on 'discovering' it in the Governor-general's menagerie in India: it was extraordinary to find such a creature in Malaysia, given that its relatives lived in the Americas. The creature had been found by Thomas Horsfield, a protégé of Raffles, who published his account after the French.[31] Naming was an important business for naturalists, as for Adam; and great men at metropolitan institutions regarded it as their privilege, but might deign to name an animal or plant after its collector. Swainson supported the rights of the Parisians to do just this, against the grumbles of true Brits. In so doing, he managed to be very rude to prominent contemporaries, and the various responses make entertaining, if unedifying, reading – passions can be raised in natural history, scientists have often been prima donnas, and the leading experts in any field in the nineteenth century were often not on speaking terms. In the end, the editor (spurred on by some readers) tried to call a halt to this clash; and the last sallies came in appendices, published at the expense of the writers.

Later volumes of this entertaining and very readable journal included an attack upon Audubon by Charles Waterton, the eccentric squire of Walton Hall; later Swainson, an admirer of Audubon, was drawn also into this controversy.[32] Waterton also gave advice on how to cope when attacked by dogs or by lions; and then in the ninth volume there is a lengthy assault on the highly artificial 'quinary' system of classification, beloved of Swainson and others, in which animals were grouped in spiralling patterns of five circles, three big and three small, arranged in different levels. The journal was thus lively, and had the advantage that natural history was much more accessible than the physical sciences: readers could really join in, and their observations and queries were welcomed. This was still a collaborative kind of enterprise, with a strong feeling that the readers are a natural group, with real if sometimes irate conversations going on. It might be compared with the contemporary issues of the venerable *Gentleman's Magazine*,[33] which exuded a strong class feeling, binding country gentlemen together in a genially bumbling way. It had a miscellaneous or even desultory character, but included a plate of Exeter Hall, a report on the zoo and a description of a device for raising sunken ships, along with its obituaries, notices of books and antiquarian contributions.

Into the different world of 1864, James Samuelson and William Crookes, a journalist and a pioneer of analysis with the newly-invented spectroscope, launched *The Quarterly Journal of Science*, absorbing an Edinburgh journal but hoping for something quite new. They perceived something like two cultures, with Art favoured over Science:[34]

> Her youthful steps have always been watched with jealousy and suspicion, and instead of guidance and support, every obstacle has been thrown in her path, her grandest revelations being frequently held up to scorn and obloquy, and twisted and tortured until they were made to appear the teachings of the Evil One.... A certain amount of scientific knowledge is now absolutely necessary to men of all ranks, and forms an essential element in a liberal education. The influence of scientific discovery is becoming daily more powerful, and making itself felt ... everywhere, – in the factory or mine, in the university or schoolroom, in the world of pleasure as in the world of pain.

Few men of science were awarded the honours and recognition that went to politicians, lawyers or theologians, and it therefore needed its own special organ: a journal open to all, not just scientific readers, who may find out how and where science is going, and how this will affect their material well-being and their eternal happiness.

The editors then ranged across pure and applied science, noting especially the racism of the Anthropological Society; doubting that spontaneous generation of life ever takes place; but welcoming most activity, such as attempts to acclimatise salmon in Australia, improved boilers in steamships, steam-hammers, and the application of photography and spectroscopy to astronomy, so that the chemistry of the stars (seen by Comte as a logical impossibility) was being realised. The papers inside were a mixture of original articles, written to be accessible (like Royal Institution lectures) to educated but unspecialised readers; 'chronicles of science' which kept them up-to-date with a wide range of sciences; reviews of books and pamphlets; a full account of the meeting of the BAAS, and of other scientific societies at home and abroad; and some notes and correspondence, including the names of the twelve gold medallists in the new Science Examinations – one of them female – and their teachers. Thus the formula was not altogether different from what we have encountered elsewhere, notably perhaps in the Royal Institution's earlier *Quarterly Journal*; but, although the articles were sometimes in parts, they were more discursive than in some popularising journals, being clearly aimed at the middle class rather than artisans or mechanics.[35]

The first issue began with a survey of Britain's (finite) coal resources; and contained one highly-original paper, by William Odling, setting out a table of chemical elements strikingly similar to what we know as Dmitri Mendeleev's 'Periodic Table', that adorns the walls of chemistry lecture theatres everywhere. But the *Quarterly Journal* was not a good vehicle for such a specialised paper, and Odling, unlike Mendeleev a little later, did not seem to have appreciated the importance of what he had done. In general, there is no doubt that reading this journal would have kept an interested outsider aware of the direction in which the sciences were going, and informed a specialist about other branches of knowledge. There were taboo-breaking articles on science, politics and religion: a fierce review in the first volume of an atheistic book, and an editorial in the second deploring a recent round-robin urging that Science and Scripture were fully compatible on the one hand, and Disraeli's sneering at Darwinism, and recent Papal pronouncements against modernity, on the other:[36]

> Secure in the sense of its growing influence and irresistible progress, Science can afford to smile at the little squibs that are from time to time thrown on its path, and may safely permit its truths to be ridiculed or perverted by persons who conceive that such a course will conduce to their political popularity, or will enable them more readily to attain religious ascendancy.

This rather pompous tone is some way from the catty remarks in the *Magazine of Natural History*, and the reader gets a vision of science as a rather solemn and serious business.

At the time of volume eight, 1871, Prussia had just defeated France, vindicating its educational system and promotion of applied science in industry. Crookes alone assumed the editorship, and there was duly an article on 'War Science', and another on artillery. August Hofmann wrote some fascinating memoirs of his early years directing the Royal College of Chemistry. Piazzi Smyth, Astronomer Royal for Scotland, wrote at length on the Great Pyramid, about which he became mildly dotty. Crookes himself included papers on 'psychic force', as displayed in spiritualistic seances, which he had written and which had been turned down by the Royal Society. This was full of detailed descriptions of what seemed to be going on, with testimony from respectable participant observers. Clearly, there was both mainstream and fringe science in a fascinating mixture, which also included an unsigned piece about a recent explosion at the factory in Stowmarket making the new and powerful explosive, gun-cotton;

and a thoughtful discussion of patent rights.[37] We note how the index in the volumes expands or contracts to fill the pages available at the end. By now, the publishers had got the bright idea of charging authors for offprints:[38]

> Authors of ORIGINAL PAPERS wishing REPRINTS for private circulation may have them on application to the Printer of the Journal, 3, Horse-Shoe Court, Ludgate Hill, E.C., at a fixed charge of 30s [£1·50] per sheet per 100 copies, including COLOURED WRAPPER and TITLE-PAGE; *but such Reprints will not be delivered till* ONE MONTH *after publication of the Number containing their paper, and the Reprints must be ordered when the proof is returned.*

To get modern values, we should have to multiply by fifty or sixty; when the price is indeed about what modern journals might charge, though many (if they still do offprints on paper, rather than electronically) supply twenty-five or so free.

After 1878, the journal changed over to monthly publication: this was a period in which the elderly quarterlies were being challenged by new and more lively monthlies, and a general speeding up of society was perceived. Crookes believed that science was moving so fast that:[39]

> intervals of three months are too long for a journal which seeks to be the connecting link between the scientific investigator and the educated reading public. A discovery made, a theory propounded, or a book published in January has, before April, been duly discussed, and new questions have attracted general attention.

The volume was much fatter than its predecessor: 840 pages instead of 575; but by 1880 Crookes' name had disappeared from the title page, and in 1885 publication stopped. In the high-speed world, weekly publication was appropriate.

Crookes himself, in a distinguished career as scientist and journalist that culminated in the Presidency of the Royal Society,[40] brought out the more narrowly focused *Chemical News* weekly from 1860, in columns and small type. At this point chemists were divided, largely between those in academe and in industry. The Chemical Society split: some members wanted a society devoted to research and publication, others a professional body, interested in fees, qualifications and careers. This latter group formed the Institute of Chemistry, later Royal; and the two only came together as the Royal Society of Chemistry after a century apart. Crookes' journal was for both: he never had an academic post, doing his important research in his private laboratory where he also undertook analyses for a fee, employing assistants whose research belonged to him. His idea of weekly publication in magazine format was taken up by another self-made man of science, Norman Lockyer, who persuaded Macmillan to launch *Nature* in 1869.

For many years, *Nature* ran at a loss, but it brought prestige to its publisher and helped build up his list of textbooks at a time when science was belatedly entering the curriculum. A distinguished list of contributors brought it authority, and it promised very rapid publication (which would be followed by lengthy papers in learned journals, with careful peer-reviewing). Gradually it displaced its rivals, achieving (as other journals had done earlier) a position in which it was read both by active researchers anxious to keep up across the spectrum of the sciences, and outsiders interested in what was happening.[41] It gradually became an institution, recognised as the forum in which scientific controversy might be carried on, pecking-orders maintained or transformed, and to which disputes might be referred: people wondered how they had managed without it. Unlike most popular science by the late nineteenth century, *Nature* was highly respectable.

Something more frivolous was also necessary. In 1865 *[Hardwicke's] Science-Gossip* had been founded as 'a monthly medium of interchange and gossip for students and lovers of nature'. It was published by Chatto and Windus, and cost fourpence. Its paper cover featured an elaborate border illustrating the seasons, surmounted by an owl presumably indicating night, and flanked by telescopes and chemical apparatus. At the bottom of the page there is a pond, where a small trout leaping to catch a fly is itself caught simultaneously by a pike and a kingfisher. In 1894, the entomologist John Carrington replaced the autodidact John Ellor Taylor as editor, and a new series began with the cover-design printed in green, on shinier paper, less cluttered in the middle with type, and published by Simpkin Marshall. It contained advertisements, which are often great fun, and inserts – promoting Pear's soap, among other things – but was respectable enough to be taken by the Royal Microscopical Society. Its staple was natural history, and there is little of what we would think of as gossip, meaning scandal. Printed in double columns, it appeared on the twenty-fifth of each month; during 1895 the price rose to sixpence.

In his preface to the new series,[42] Carpenter declared that he aimed at five classes of readers, from the amateurs or dilettanti, through field-collectors (men and women) who provide material to be worked out by specialists, the third class; and on to Philosophers of Science, like Darwin, Weismann and Tyndall (we might note that Darwin and Tyndall were dead); finally reaching a fifth class, happily small, of snooty and dogmatic 'Pharisees of Science' who look down cynically upon mere collectors. All these would benefit from the journal, which would help beginners by answering queries and naming specimens, and also indicating interesting collecting-grounds. It would report too upon meetings of societies, and interesting facts or exhibits there; facilitate the exchange of specimens, books and instruments; and review books and instruments, being positive and letting the unworthy sink into unnoticed oblivion. A monthly feature would deal with foreign science. All this was not very different from the previous formula; where the 'gossip' had meant cheery anecdotes about being found in awkward positions when butterfly-hunting or travelling by train with escape-prone creatures.

In late 1896, the back cover of several issues was filled with an advertisement for Cadbury's cocoa, showing a bearded scientist with chemical apparatus and a microscope analysing it – and no doubt establishing that there were 'no chemicals used'. In October 1898, there were articles on natural history in Ireland, on the persistence of lowly creatures in an evolving world, on land shells, on insect-evolution, on infusoria, on Lancashire botany, on Maltese caves and their denizens, and (for children) on microbes. After a brief report of the BAAS, we find diagrams of apparatus for transmitting and receiving wireless communications – really up-to-date technology. Astronomical information occupies a page, and 'science gossip' another. That includes the news that the National Portrait Gallery was waiving its usual ten-year rule in order to add Thomas Henry Huxley; a report of 'nonsense' in daily papers about danger from sharks to bathers in English waters, and of a caterpillar scare; and gloomy thoughts on Tyndall's hobby:[43]

> There seems to be a fatal association of tastes for scientific research and Alpine climbing. The terrible accident which occurred to that distinguished electrician, Dr. John Hopkinson, D.Sc., F.R.S., when he, his son and two daughters were killed, is by no means the only case where a well-known man of high scientific reputation has lost his life through mountain climbing. The list was, unfortunately, already a long one.

It may be that the macho elements of mountaineering were really attractive to scientists, whose work might otherwise seem sedentary and unadventurous: but other intellectuals too, as memorials in the church in Zermatt show, were drawn to hazardous activity in the Alps. There were in the journal also reports on geology, and on scientific societies, brief book-reviews, and advertisements for apparatus, and for the London Zoo.

Science-Gossip entered the twentieth century in the same vein. In December 1901 natural history was very much the mainstay, but there were two-and-a-half pages on astronomy, another science in which the amateur could play an important role; and a page each on chemistry and physics, reporting the possible presence of selenium in some beers, and describing the Zeeman effect in spectroscopy. Photography, and optical apparatus – not only microscopes but the prismatic binoculars which were to revolutionise the study of animal behaviour – were major areas of interest. In February 1902 there were notes on the smallpox epidemic, where those who had neglected to be vaccinated, or objected to vaccination, were at vastly greater risk: in consequence, the lymph was now in greater demand. There had also been a meeting of science teachers, now concerned with:[44]

> the giving of a training, and not merely the imparting of a series of facts. This, by the bye, with all due deference to the old-fashioned schoolmaster, is not worthy of being considered an education, nor, must we own, is a knowledge of scientific method without general culture.

The world had changed a great deal since the early days of *Nicholson's Journal*. Most science had become an elite activity, with highly-qualified professionals working in well-equipped laboratories, and prone to look down upon amateurs with more or less indulgence. Nicholson, Tilloch and Thomson, and also Banks, could think of all who read their journals as part of the army of science advancing upon ignorance, some with more time and aptitude than others. Gradually this army came to look more like others, with officers, NCOs and men, corresponding roughly to the 'classes' aimed at by *Science-Gossip*. The scope for low-ranking soldiers and their initiative was reduced. The army of science, again like others, was expensive, especially as it became a professional body rather than a militia, needing both expensive training and the latest weaponry, in the form of expensive apparatus.

This training, scientific education, was something that had preoccupied Huxley, lecturing to students, teachers and working men, and serving on the London School Board. Important questions were how to get facts across in coherent and memorable form; and also what scientific method (or perhaps methods) entailed, and how applicable scientific thinking was in other spheres. Now it is time to look at the implications of suspending judgement, and the rise of the scientific sage.

11 Suspending judgement

Scientific method seemed to many in the nineteenth century to be somehow the essence of science. Learning science was hard: lots of facts, difficult language, symbols and concepts, and often all of very remote relevance to ordinary life. Only some people were ever going to need to master it; but perhaps others could pick up the scientists' method of finding truth and apply it elsewhere, so that their conclusions were not dominated by emotion, deference to authority, or mere habit. The motto of the Royal Society was, and is, 'Nullius in Verba', take nobody's word for it. Careful testing was required, though it could be ambiguous. It was said that a spider could not escape from a circle of unicorn's horn, and the early Fellows duly tried the experiment. The spider got away, which showed either that the claim was false, or that their unicorn's horn was not genuine. That kind of difficulty attends all experiment, though in practice scientists try to control the circumstances so that only one option is really at all likely, and the result therefore convincing. Francis Bacon's hope was that we could rid ourselves of all our 'idols' or preconceptions, but that seems optimistic; science does rest upon some metaphysical assumptions to do with uniformity, which is why psychical research is so problematic – telepathy, even when no trickery is apparent, seems to have good and bad days.

Huxley loved lecturing to working men, as did other Victorian men of science: there was no syllabus to get through for reluctant hearers with an exam to pass, and no effete and bored persons who had drifted along because it was the thing to do.[1] These audiences had done a hard day's work, and were keen to broaden their knowledge. They shared with Priestley the hope of the Enlightenment, that science would set them free and empower them. They were almost certainly not going to be able to contribute to the edifice of science. Bacon believed that everyone could add his brick, but by this time there were barriers which made it very difficult for the uneducated to get into real science. Huxley himself sought to establish a well-trained corps of professional scientists, excluding the amateur enthusiast. His science was museum palaeontology and laboratory physiology. Only in astronomy and natural history was there serious scope for the amateur, and an astronomer had to own a telescope – well beyond the means of a working man. Country parsons and artisans could indeed study plants and animals in their habitats, and their work could be of real interest and importance for those like Hewett Cottrell Watson[2] and Charles Darwin, concerned with the habits, distribution and variation of organisms. This was not, however, at that time very interesting to professional biologists experimenting in laboratories or classifying in museums. So those working men came to Huxley to pick up his enthusiasm, to wonder at the transformations of creatures over time, and to learn how his conclusions were reached.[3]

Huxley sought to demonstrate that science was trained and organised common sense. He showed them that they were following scientific method in their ordinary lives whenever they reasoned from evidence back to causes. For him, as for Bacon, it was not a matter of geniuses like Newton having extraordinary inspirations: there was more perspiration – hard work, suspended judgement, careful testing, and cautious generalisation. There could not always be proof, in the way that geometry was proved. Faced with different kinds of crocodiles going back over millions of years, his audience could suppose with him that recent ones had evolved from earlier forms, or could infer that God at intervals killed off one kind and created another, not very different. Just as in a law court, the verdict has to be beyond reasonable doubt, so it was in science: probability, as Bishop Joseph Butler wrote in the eighteenth century, was the very guide to life.[4]

For Huxley, the adjectives 'trained and organised' were all important. We all think rather scientifically in ordinary life, but to do it properly entailed for him a scientific education with practical training, as he himself had in medicine. Without it, nobody would know where to start or how to proceed. Also, science requires organisation, institutions. Schools and colleges with their laboratories, specialised societies and more open ones with their meetings, conferences and lectures, academies like the Royal Society bringing together the most eminent practitioners, publishers to bring out books and journals, industry alive to new knowledge – all were and are required to keep the scientific enterprise moving. Huxley worked and taught at the School of Mines and its museum, and then at the Normal School in South Kensington where science teachers were trained. He had been opposed (partly because Owen wanted it) to moving the Natural History collections from the British Museum all the way out from central London to South Kensington, but his career took him there too, in institutions that, after his death, came together to form Imperial College of Science and Technology.

Descartes, whose philosophy was based upon doubt and its resolution, was one of Huxley's heroes; his *Discourse of a Method*, written to accompany his treatises on geometry, optics and meteorology, was translated into English in 1649.[5] Thereafter, there were available two alternative accounts of scientific method: his mathematical model, where deductions from hypotheses were tested by reason and experiment, and Bacon's, where generalisations were cautiously made from well-ascertained facts. Many others joined in. When Davy and Wollaston died within a year of each other, they were duly compared: Davy ardent, enthusiastic, leaping to hypotheses in seeking truth; Wollaston cautious, scrupulous in analyses, careful to avoid error.[6] We should not take these too seriously, but they do illustrate two scientific types. Davy's friend Coleridge wrote on method, which for him was a way of recognising what was important in things and relationships, a criterion of excellence in authorship.[7] In his journal, *The Friend*, he included an essay on method, composed as the introduction to the *Encyclopaedia Metropolitana*, but garbled when published there.[8] Method was:

> a distinct science, the immediate offspring of philosophy, and the link or *mordant* by which philosophy becomes scientific, and the sciences philosophical.

Coleridge took examples from botany and chemistry to show what should be done, and emphasised as we saw the role of imagination. Socrates for him had been the founder of inductive philosophy, and Bacon was not just a cautious empiricist but stood in the Platonic tradition. An admirer of the great surgeon John Hunter, and well-informed about medicine, Coleridge became disenchanted by Davy, preoccupied as he saw it by little things and analytic methods. He wrote of the imperial ambitions of chemists, practising what they wrongly saw as the fundamental science, in the *Friend* and in his posthumously published magnum opus, the slim *Theory of Life*:[9]

> The new path, thus brilliantly opened, became the common road to all departments of knowledge; and to this moment, it has been pursued with an eagerness and almost epidemic enthusiasm which, scarcely less than its political revolutions, characterise the spirit of the age. Many and inauspicious have been the invasions and inroads of this new conqueror into the rightful territories of other sciences; and strange alterations have been made in less harmless points than those of terminology, in homage to an art unsettled, in the ferment of imperfect discoveries, and either without a theory, or with a theory maintained only by composition and compromise.

The chemical method of analysis and synthesis, getting down to atomic composition, could be reductive and materialistic when applied to medicine, as to Coleridge and others it seemed to be in the lectures of William Lawrence,[10] censured by his colleague John Abernethy for blasphemy.[11] Discussions of method could be explosive, bringing debates out from academe into the public domain. Accusations of improper method have always been a standby for conservatives: later in the century they were levelled by critics such as Wilberforce against Charles Darwin. Because science depends upon proof, strict and geometrical or inductive and quasi-legal, a close look at method seems very reasonable; but just as Coleridge's poetry showed how ballads and conversation-pieces could do something very new, so great scientists have extended current ideas of method, and broken the rules that bound their predecessors.

Coleridge became the Sage of Highgate, his monologues listened to with more or less patience, and his table-talk recorded for posterity. Davy knew that he was under a death sentence when he wrote *Consolations in Travel*, but that book was in effect a claim to be a scientific sage, whose ponderings on time and change must be of general interest, and indeed his legacy.[12] Had he lived, he would happily have become a sage – as Coleridge and Thomas Carlyle demonstrated, a happy marriage was not an essential part of the role. Coleridge's remarks indicate that chemistry, perhaps the key to the new scientific revolution that spread from France, was an intellectual ferment comparable in influence and popularity to the political ideas that also rippled out from Paris. Public understanding of science, and especially of its method, was therefore important. Becoming a scientific sage was a reasonable objective: and the propagation of scientific method was a part of it. John Herschel and Huxley duly became such figures.

Herschel's book on method, *Preliminary Discourse*, (1830), in *Lardner's Cabinet Cyclopedia*, became very influential because of the 'Newtonian' inductive philosophy it imparted, which seemed in line with the new science of the day; and because it was accessibly written.[13] Whereas Bacon had been suspicious of mathematics, Herschel had a Cambridge mathematical education and had been one of the group that modernised it, introducing the new Continental notation and methods of mathematical analysis there. Aware therefore of up-to-date French and German scientific work, he saw mathematics as the gateway to physical science. He admired Bacon; but his method was thus different from the careful experimental generalising which Bacon was generally taken to have advocated. But he also avoided speculative system-making through insisting on Newton's idea that any proposed explanation must involve a *vera causa*, a genuine and plausible mechanism analogous to other known causes.

Following his voyage to South Africa to observe the southern stars, he was a hero.[14] He was made a baronet, in due course was appointed to the post Newton had held, Master of the Mint, and was consulted by everyone wanting a learned opinion on a scientific question. Eventually (as Charles Darwin was to do) he bought a big house in Kent, remote from the railways, and settled down in relative peace with his wife, twelve children and fifteen servants. Distinguished science, and writing on method, had brought him a reputation: a good artist, he was a pioneer of photography (though not himself a photographer, not needing to be) – Margaret Cameron's photograph of him is a classic portrait of a seer. His *Preliminary Discourse* stimulated two other prominent thinkers, William Whewell and John Stuart Mill, to publish their versions of scientific method.

Whewell, Master of Trinity College, a carpenter's son who became one of the great know-alls and pundits of early-Victorian Cambridge, had been Professor of Mineralogy and then of Moral Philosophy.[15] He was a great admirer of mathematical physics, and believed that Bacon's inductive philosophy had been wrongly interpreted as data-collecting and generalising. For him, getting the right end of the stick was essential. Each science had its fundamental idea (symmetry in mineralogy and crystallography, likeness in taxonomy) and until this general principle was intuited, the facts that would bring it into clearer focus could not appropriately be collected. He reviewed Herschel's book in the *Quarterly Review*, praising his friend's work but emphasising 'colligation', finding the concept that forms the bond of unity holding phenomena together. This is not remote from Coleridge's Imagination, and was much closer to Idealism than Herschel's position had been. As a cleric, Whewell was also keen to argue for the importance of the moral and religious beliefs of the scientist (his word) in finding truth. Whewell was impressed by the consilience of inductions, when distinct lines of inquiry converge upon a point. Such convergence is indeed important in confirming the likelihood of wide-ranging theories, and featured in the *Origin of Species* which can be taken as an example of Whewell's method: though that thought was repugnant to Whewell, who would not have the book in the college library.

Whewell did not go all the way with the hypothetico-deductive approach to science: in writing his *Bridgewater Treatise* he had been appalled by the arrogance of pure mathematicians (notably Laplace) seeking to impose patterns upon nature. Just as Coleridge distinguished imagination from fancy, so Whewell separated the search for truth from finding equations that worked, saving the phenomena, but not *verae causae*. Pure mathematics in Whewell's view should be kept in its place: it led too readily to atheism and overconfidence. Applied mathematics he saw as the key to Cambridge education: the way the world is, the sullen facts,[16] were ever before the eyes of the students. Scientists were engaged in the serious business of finding out how it really worked; their theories were a representation of reality. As George Gabriel Stokes put it:[17]

> A well-established theory is not a mere aid to the memory, but it professes to make us acquainted with the real processes of nature in producing observed phenomena.

Whewell's books on method, the *History* and the *Philosophy of the Inductive Sciences* (1837, 1840) are substantial multi-volume tomes, which give a wonderful conspectus, as a powerful intellect grasped with the whole range of the sciences of the day, finding and imposing order. Aimed at the well-educated and well-to-do, they made a serious contribution to that public's understanding. In fact, while to our generation a formidable read, they are a wonderful example of the method that Whewell favoured, inaugurating the history of scientific ideas. The problem turned out to be that the sciences have changed their focus and direction, undergoing revolutions and experiencing new syntheses, and that Whewell's categories did not last. But he opened up discussions of method, bringing a greater realism: in Britain particularly, scientists had often claimed to be working by simple induction, when in fact they were imaginative, intuitive – honesty in discussion of method is a great thing. The nineteenth century was a great age of literary criticism, denouncing feeble writing but promoting good work and seeking to understand what made it good – and Whewell might be best seen as a kind of scientific critic, with the same sympathetic seriousness and underlying passion – certainly not against science any more than Matthew Arnold was against literature.[18]

Whewell deplored utilitarian ethical calculations as leading to atheism. Mill, as an agnostic, utilitarian and correspondent of Comte, was therefore on the opposite side; and, in his *System of Logic* (1843), he presented a view of scientific method as empirical generalisation. Mill was not, like Whewell, himself a prominent scientist, and did not have an academic career. But in his own time, his empiricism and his liberal political views made him much more influential with the general public than Whewell ever was. Moreover, in Oxford after the burst of religious enthusiasm kindled by the Tractarians in 1833 died down when Newman became a Roman Catholic in 1845, Mill became the most admired living philosopher. He was associated with the *Westminster Review*, which was progressive and opposed to organised religion; science and its empirical method seemed to fit the bill as a basis for this kind of modern thinking. Darwin was anxious to know what Mill would make of the *Origin of Species*, and was delighted to hear that 'he considers that your reasoning throughout is in the most exact concordance with the strict principles of Logic'.[19] Darwin, who believed that all observation must be for or against some view if it is to be any use, was much relieved; but it is curious that Mill, more empirical then Herschel or especially Whewell, should have been happier with the book than either of the others.

One of those who responded to Mill and Whewell was Baden Powell, Professor of Mathematics at Oxford, liberal theologian, and father of the founder of the Boy Scouts. Upset by Tractarian enthusiasm for religious controversy and what he saw as obscurantism, which led to decreasing interest in science in the 1830s and 1840s, and coming to believe that religion and science must be firmly separated, he turned to writing on scientific method.[20] He despaired, prematurely as it turned out, of the prospects of science in Oxford, whose graduates would lack the understanding he could have provided. In his *Essays* of 1855, he urged the unity of the sciences, and the order and harmony of the world.[21] He delighted in Lyell's vision of slow change over millions of years, and when the *Origin of Species* was published he was among the clergy who welcomed it, in his contribution to the deeply-shocking *Essays and Reviews* (1860).[22]

By inductive sciences Powell, like Whewell, meant empirical disciplines, sometimes then called 'mixed sciences' and distinguished from mathematics and logic, which were purely deductive and might or might not be applicable to the real world. He:[23]

> considered some instances in which the discoveries of science are undeniably at variance with doctrines which had become identified with popular belief, or had even been erroneously received as part of the established creed.

Deploring scientists who concealed such divergences in an ambiguous phraseology, and defenders of religion who became 'followers in the train of prejudice rather than its correctors and enlighteners', he urged a liberal view of religion where the empirical claims of science must be allowed and the Bible interpreted accordingly. He repeatedly cited the essays, *The Soul in Nature*, of Hans Christian Oersted:[24]

the philosophical views broached in a posthumous work, which has so fitly and honourably crowned the labours of the great Œrsted, and added a new claim to our admiration of his genius. In those essays he maintains the proposition that 'the laws of nature are the same as the thoughts within us'.

Oersted had become famous with his discovery of electromagnetism in 1820; before that, his association with Johann Ritter and others in German Naturphilosophie had made him suspect in Britain as a metaphysician rather than a man of science.[25] Herschel shared Powell's admiration, declaring when welcoming Oersted to the BAAS in 1836 that he wished all would give up 'the hasty generalisations that marked the present age' and adopt Oersted's 'more safe system'.[26] After making his great discovery, Oersted played little part in the development of electromagnetism, where Andre-Marie Ampère and Faraday became dominant; but in his native Denmark, a small agricultural country, he played a great role as scientific guru. There he found himself writing for, and lecturing to, a wide range of audiences using a language few outside Scandinavia understood; and in a particular religious situation. One of the great Danes of the past had been Tycho Brahe, the sixteenth-century astronomer, whose ingenious planetary theory had the Earth at the centre, and the Sun (carrying all the planets in their orbits) circling it – thus, he believed, reconciling the best bits of the Ptolemaic and Copernican systems. Oersted sought to bring to the public an understanding of newer science, leading his compatriots into the nineteenth century; and he believed that science was fully compatible with spirituality.

His *Soul in Nature* is thus a curious collection of dialogues, poems, addresses, lectures and essays, with varied dates, contexts and audiences; but throughout there is admiration for what to most British empiricists would seem a dangerously a priori method of reasoning, and a dynamical science in which apparently solid bodies endure only as fountains do, through the Heraclitean flux of their constituent particles. It conveys fervent pantheistic spirituality rather than orthodox religion; and, like Powell, Oersted pointed out that the geological record demonstrated how (contrary to a literal reading of *Genesis*) suffering, death and destruction long antedated the appearance of humankind. The book was undoubtedly a contribution to method and world-view, rather different from those of John Stuart Mill. But before drawing back from a priori science, we should remember a remark attributed to Faraday, the great experimentalist:[27]

> Before we proceed to consider any question involving physical principles, we should set out with clear ideas of the naturally possible and impossible.

Science cannot be done without intuition, even by those supposed by contemporaries to be Baconian inductivists.

The English translation of Oersted's book did not do well, coming before a very different (and more sophisticated) public from his in Denmark; and when Darwin the deductive theorist read it, he recorded 'dreadful' in his notebook.[28] Darwin was cautious, and like many contemporaries had been deeply concerned about German Naturphilosophie, exemplified in Lorenz Oken's *Elements of Physiophilosophy*, controversially published in English translation by the Ray Society in 1847[29] (and thus for a public of keen natural historians). It begins with propositions like 'Spirit is the motion of mathematical ideas. Nature, their manifestation'; and proceeds through astronomy to geology, with various interesting taxonomic tables, and on to botany where '*the motion of the sap is imparted through the antagonism of the respiratory and digestive processes.* For these two processes are the combination of the Chemical with the Electric, which is the galvanism.' For a public anxious for a big picture, this provided an alternative based in philosophical idealism to the materialism of *Vestiges*; and transcendental anatomy, in which animals represented various realisations of some Platonic Idea, seemed a way of understanding relationships without the crudity of Chambers' evolutionary ideas. In particular, the idea that Goethe probably had first and Oken emphasised, that the skull was modified from vertebrae, seemed to unite the whole vertebrate kingdom under one common plan.[30] In 1840, Charles Eastlake, Royal Academician as well as Fellow of the Royal Society, and first Director of the National Gallery in London, translated the less polemical parts of Goethe's *Farbenlehre* in which he had taken on Newton's reductive-seeming physics of light and colours.[31] The great landscape painter, J.M.W. Turner, experimented with Goethe's ideas, in seeking to achieve perspective through colour.[32] For publics unhappy about the separation of science from humanities, there was an alternative, coming (besides the modernity of Liebig's experimental chemistry) from Germany, where the role of the active mind was much emphasised. This could take root in minds prepared by Coleridge.

Among admirers of German thought was the polymath T.H. Huxley, but his researches convinced him that the vertebral theory (upheld now by his old enemy, Owen) was wrong.[33] He also espoused a view of science, as trained and organised common sense, very different from that of Oken and Goethe. In particular, he had no time for religion, organised or even personal. Although he presented his work as a New Reformation, carrying through what Luther had begun, its basis was in suspending judgement: the word he coined for this in 1869 was 'agnosticism'. Crookes had objected to Faraday's maxim given above, because men of science ought in Baconian fashion to be completely open-minded, dismissing all preconceptions or 'idols' from their minds: then they would investigate psychic phenomena as he had done, and report just what they saw (or believed that they saw, as opponents charitably put it). As Crookes well knew when he was at work with his spectroscope, the scientist has to be guided by a world-view, and by current theories, even when opening up new territory. Faced with booming buzzing confusion, everyone has to look for a message amidst all the noise, and decide what is relevant: what, in Darwin's terms, counts for or against a view. But it is still possible and important to guard against preconceptions and prejudices, even if such things cannot be eliminated; and, for Huxley, that was the beginning of science.

He had abandoned organised religion as a medical student, aware like other radicals in that overcrowded profession of how the established powers of church and state were keeping the poor down, rather than preaching the gospel to them by actions that would speak louder than words.[34] On his return from his tour of duty as assistant surgeon aboard HMS *Rattlesnake*, he endeavoured to make a scientific career for himself; finding that the clerical stronghold on the universities of Oxford and Cambridge kept people like him out. Convinced that science was a vocation and should be a profession, rather than a part-time activity for clergymen whose first loyalty was to Christian doctrine, he supported secular education – though believing that the Bible should feature in it, as a source of poems, stories and moral teachings crucial to culture. He was a great admirer of Old Testament prophets, who boldly stood up to kings and crowds, telling them where they were in the wrong. He saw himself as a prophet in that sense, calling for appreciation of science. His own lecturing and writing gained power from the Biblical references and echoes that were an important part of his style (and that of contemporaries), highly acceptable, along with tags from Shakespeare and Tennyson, to Victorian publics, indicating seriousness and breadth. Rather than seeing the fear of the Lord as the beginning of wisdom,[35] though, he insisted that we be honest about what we can and cannot know. That, for him, was a clear message coming out of the success of scientific method.

It is curious that agnosticism had its roots not only in the positivism of Comte, but also in the theological writings of Henry Mansel.[36] In his *Limits of Religious Thought*, published in 1858 by John Murray, later to publish the *Origin of Species* (1859), he revived the *via negativa* in the Church of England. In an old tradition, he argued that positive statements about God must be false, because our finite minds cannot grasp the infinite: it is by denying that God is like this, or like that, that we can begin to fathom Him. Mansel sought to save dogmatic theology from dogmatism. The book was a success, with a fourth edition by 1859. In that year, Mansel was elected the first Wayneflete professor of moral and metaphysical philosophy at Oxford, and in 1868 was chosen Dean of St Paul's Cathedral in London. It was however outside the churches that God became the Unknowable or the Absolute – rather like the First Cause for eighteenth-century Deists.

For Huxley, nothing definite could be said about the unknowable. Mansel's use of doubt as the foundation of belief seemed absurd. He had no time for the dogmatism of atheists or of most theologians, and as we saw he regarded the positive religion of humanity propagated by Comte as ridiculous: Catholicism minus Christianity.[37] Agnosticism was a method, not a creed. What was necessary was to work at questions which had definite answers. Huxley's philosophy of science, not at all far from that of Mill, was developed in controversy with others, and is relatively informal and accessible, found in his popular writings. Like Mill, he had begun by admiring Comte, but increasingly saw the need to separate himself from him: especially because he was often bracketed among positivists, a significant group in Britain and the USA after Harriet Martineau had condensed and translated his *Positive Philosophy* in 1853.[38] The book was published by John Chapman, of the *Westminster Review*, whose seamy life had made him unrespectable. Positivism could be seen in the light atheism had been, as an excuse for immorality. Huxley would have none of that. Morality was not attached to organised religion, and agnostics should lead lives as respectable as anyone – indeed they might be less tempted to hypocrisy than churchgoers.

Huxley did not accept Comte's three stages: children did not move from theological to metaphysical to positive thinking – all three co-existed and always had, and the human race would never reach the positive stage.[39] Huxley also strongly criticised Comte's science: he had believed in phrenology, written that biology was a matter of observation rather than experiment, and (like the author of *Vestiges*) indulged in inaccuracy, inconsistency and misinterpretation. He disliked Comte's meddling systematisation and regulation, as he saw it: Comte was a prescriptive philosopher of science, laying down rules for doing it properly. Huxley on the other hand was a scientist generalising from his own experience and that of others to produce a descriptive philosophy based upon excellent practice; and was strongly averse to being told what to do or think. This he saw as essential to science.

Huxley's hero was David Hume, the sceptical empiricist. He was not, in the middle years of the nineteenth century, the canonical figure he has become, featuring strongly in any history of philosophy: he was known to the public as an historian, a religious sceptic who had proposed some counter-intuitive ideas about causation, and had alarming thoughts about miracles. In 1874–5, there was a new edition of his writings, and philosophical idealists like T.H. Green of Balliol College, bringing Kant before British students in place of Mill, were interested in the man who had disturbed Kant's dogmatic slumbers and set him off on his 'Copernican revolution'. In 1878 Huxley published a book about Hume, which was reprinted six times before his death in 1895.[40] This was part of a series, edited by John Morley, on English Men of Letters; Huxley toyed with editing a series on English Men of Science, and joked that in writing the biographical part of this book he had made his nearest approach to fiction. Certainly it told readers as much about its author as its subject.

Psychology was Huxley's, and he believed it had been Hume's, route into philosophy. Method in psychology should be the same as in other sciences, based upon experiment, and hypotheses to be tested and criticised until such scaffolding can be taken down. Huxley advocated 'mitigated scepticism' as a recipe for keeping dreaded dogmatism and superstition at bay; and believed that operations of the mind are functions of the brain – a notion he thought as compatible with idealism as with materialism. He was a firmer believer in law and order in the world than Hume had been: indeed, blank misgivings about laws of nature would inhibit the serious pursuit of science. But sceptical mitigated empiricism was right for science as for life; and his enthusiasm for Hume went with his recognition that we all do science whenever we look for the causes of things.

Huxley was an amateur philosopher, reading widely often in order to demolish a foe; but he was a teacher's son, and a keen teacher for most of his life. From the 1860s, his role in education and in popularising was central, and his personal research took second place: his greatest pupil, Michael Foster, declared that Huxley:[41]

To a large extent deserted scientific research and forsook the joys which it might bring to himself, in order that he might secure for others that full freedom of inquiry which is the necessary condition for the advance of natural knowledge.

At South Kensington, he taught students his method, inaugurating laboratory teaching and research in zoology and physiology for his students. By the time he died, in 1895, many professors of biology in Britain had worked under him, and his disciplines had become secular professions distinct from medicine and natural history. His campaign to replace clergy in the learned world by meritocrats, expert in their field, and aristocratic patrons by powerful professors, had succeeded. In 1870, when compulsory education at last came to England, he stood for the London School Board, and was elected. He did his best, there and more generally, to ensure that some science got into elementary education.

He was also a great public figure, sitting on Royal Commissions whose reports to Parliament led to legislation. An important one, where he both heard and gave evidence, was the 'Devonshire' Commission (so-called from its chairman, the Duke of Devonshire, a great patron of science) on scientific instruction and the advancement of science. Their eight substantial and closely-printed reports, published in 1872–5, included evidence given, plans of teaching laboratories, and conclusions – which were that more science should be taught, in secondary schools and in universities. Such public inquiries, the documents they produced, and the debate they generated were and are very important in public understanding of science.

This one, set up in 1870, had Norman Lockyer, civil servant, astronomer and founder-editor of *Nature* (1869), as its Secretary: its members, a distinguished group, were the Duke; the Marquess of Landsdowne (a descendent of Priestley's patron, Lord Shelburne); Darwin's friend, John Lubbock; James Kay-Shuttleworth, eminent in public health and education; Bernhard Samuelson, industrialist and educationalist; William Sharpey, physiologist, London medical professor and a long-term Secretary of the Royal Society; Huxley; William Allen Miller, physical chemist and spectroscopist who had identified chemical elements in the stars, and George Gabriel Stokes, devout Cambridge mathematician, editor of the Royal Society's *Philosophical Transactions*.[42] Miller died in 1871, and Henry Smith, mathematician and, from 1874, Keeper of the Museum in Oxford, at that point one of the best science centres in the world,[43] joined the commission. They commended the core curriculum drawn up by the London School Board, including systematic 'object lessons' embracing elementary physical science, and described and tabulated the science teaching and examining done under the aegis of the Science and Arts Department in the new system of 'payment by results'. The number of schools involved in this programme had grown between 1860 and 1870, from 9 to 799, and the number of those under instruction from 500 to 34,283. Lectures were given by scientists, including Huxley, to teachers, who were given 'pecuniary assistance' to attend.

But it was secondary and higher education that was the commission's brief.

Between 14 June 1870 and 11 July 1871, evidence was taken from ninety-five witnesses (including some, like Henry Cole, who came forward twice). Their answers, set out in 9,261 numbered paragraphs, occupied 629 large pages in double columns. There are interesting exchanges with, for example, Isaac Lothian Bell and Sir William Armstrong, prominent industrialists in north-east England, and the last witnesses, about the recent foundation of a College of Physical Science in Newcastle upon Tyne: both were strongly in favour, indicating a change from a previous generation when employers had scorned engineering qualifications from the nearby University of Durham, under whose auspices the College was beginning. Bell cited Davy's safety lamp as an example of what science could do for industry. Pressed, Armstrong feared that 'young men at the Universities, at all events, are liable to acquire fastidious habits that do in some degree disqualify them for the rough life that they would have to lead in a workshop'; but he remained optimistic, and suggested that some government funding would greatly benefit the new institution.

This was revolutionary idea: state support for, and control of, higher education was normal in France and Germany, and land grant colleges (the future state universities) were coming in the USA. But in Britain for many years the only institution to get a grant was the Roman Catholic seminary at Maynooth in Ireland, in the hope that its graduates would be pro-British. Not until the late 1880s, when William Ramsay the chemist was Principal of University College, Bristol, did Parliament vote money as Bell and Armstrong had hoped it would, for the new institutions with their strong and expensive science base, saving them from collapse.[44] Public enthusiasm for science in places like Bristol was by no means universal, especially if it was expensive.

Earlier in the report, there had been a great deal about Oxford and Cambridge, which in the wake of various commissions and reports were by this time offering degrees in scientific subjects; much of the evidence is a fascinating exhibition of academic conservatism. The value and importance, to the individual and to the nation, of scientific education was reluctantly recognised by crusty dons devoted to a liberal education common to all graduates, and fearing a fragmented, over-specialised world of 'two cultures' or more. But it is also of great interest to see slow adaptation going on, under outside pressure, in institutions which thirty or forty years before had been preparing the majority of their students for ordination in the Church of England, where the teaching had been in colleges rather than university-wide, and where the established teaching posts were often assigned, it seemed irrevocably, to classics or mathematics. We might note that similar processes were beginning in the USA, as Ivy League universities underwent similar transitions. There were doubts expressed about the place and value of examinations, and the swotting they encouraged. The recognition of research as crucial in academic life came in with the sciences, into universities which had been largely teaching institutions with what came to be seen as narrow and ossified syllabuses.

The index of the volumes contains one entry under 'women': University College, London, in a revised charter approved by Parliament in 1869, had acquired the power of instructing women as well as men. Other universities were to follow this example, some of them exceedingly slowly. The second volume contains eight separately paginated reports on different institutions, with appendices; and the third, slimmer and often bound with the second, sixty pages of further evidence from witnesses (taking us to paragraph 14,553), correspondence and the index. Scientists used the commission to promote their views and interests, sometimes controversial, as when Kew hoped to take over the herbaria housed in the British Museum; an idea the commission turned down.[45] There were still very few universities in England (the founding of most provincial 'redbrick' universities came soon afterwards, and they were followed by polytechnics in the latter years of the century), but we learn about what went on in London and Manchester and Durham/Newcastle; in the Scottish universities, and in Dublin, Belfast, Cork and Galway in Ireland; in 'public' and 'grammar' schools; in Mechanics' Institutes; and in all sorts of other places. There are syllabuses, catalogues of apparatus with costs, timetables, reports of laboratory teaching and examination processes, and booklists.

The most striking feature of the second volume is the illustration of school and college botanic gardens, observatories, museums, laboratories and lecture theatres, with detailed descriptions of their layout: steps and pathways, domes, cupboards, shelves, gas and water piping, and draught- or fume-cupboards. There are plans, and also illustrations of the best examples: that for University College School's laboratory shows apron-wearing boys, and the bank of benches so that the room can be used for demonstration-lectures as well as practical work. Given that there is a table at the end of the volume listing responses to the questionnaires sent out to 205 endowed schools in 1871, and again in 1874 to the 128 who had replied (87 responded), it is clear that many schools were in need of such examples if they were to emulate those teaching science seriously. There was more going on in 1874, but a number of the schools that responded had none, and presumably those who did not respond were in the same position.

Formal teaching of science should have led to greater public understanding, but it could mean that those attracted to science got their own classes and others could take no notice. Willy-nilly, the world was becoming more specialised; and the British system became particularly specialised. Chemistry was the science most studied in the late nineteenth century, as Britain, France and the USA followed Germany in teaching and research. It was, after all, very useful, both in industry and academically in medical sciences, geology and branches of physics, even though it no longer seemed fundamental in the way it had to Coleridge's generation. Its experimental method gave it some of the aspect of a craft, though allied now to theory with molecular models and early classification tables for the elements. But physics was not far behind it in popularity. Tyndall, Huxley's ally in the X-club, had in his Belfast Address appealed to atomic theory, energetics and evolution as the new big picture, in which there was no place for organised religion. Agnosticism seemed to go with scientific method, firmly controlled by evidence while including imaginative and intuitive thinking, and respect for the sublime. But Tyndall was out of step with many of his eminent physicist contemporaries, including Stokes, for whom religion rightly understood was an ally of science. They were building the great edifice of classical physics: and it is to this newly-fundamental science, with some features very open to public understanding, that we now turn. It was closely allied to new technology, had large-scale implications, and was by the end of the century upset in an intellectual revolution – all of which made it fascinating.

12 Classical physics

In the first half of the nineteenth century, there had been many separate sciences, little related to one another, which came under the umbrella of natural philosophy. Sometimes some of them had been and were grouped under the heading 'physics' (coming from the Greek word for nature):[1] but this label might exclude mechanics, and include meteorology and even magic, and corresponds very weakly to our notion of what physics is. The labels given to sciences, and the boundaries drawn between them, are matters of convenience, social constructions rather than discoveries about the world; but, because they are taken seriously, and have definite effects, they can indicate understanding by relevant publics at particular times and places. Careers, circles of friends and colleagues, learned and popular societies, and publications all depend upon how sciences and their frontiers are perceived. Different taxonomies put different disciplines at the top or bottom of the scale in a pecking order; psychology or sociology may be perceived as the apex of science because its subject matter is so difficult, or particle physics because it is so basic. Such a hierarchy was prominent in the writings of Comte, who despite his mathematical training, saw sociology (a name he invented) as the highest science.[2]

Classical physics was an international development, beginning, like so much else, in Napoleonic France: with Siméon Poisson's work in applied mathematics, and Jean-Baptiste Fourier's study of heat-flow.[3] This had little public impact, being highly mathematical and unabashed about it. But then Sadi Carnot imagined an ideal, reversible heat engine, and used this theoretical model as a basis for understanding how steam engines worked and might be improved, presenting it in accessible form in a little book meant to be popular.[4] This attempt to achieve public understanding was ignored by everyone for two decades: the others were influential in the tiny communities who could follow them. When interest was revived, France was no longer the pre-eminent power in science that it had been; and it was in German and British contexts that the science of energy evolved. International rivalry and claims for priority then, as often in science, fuelled interest in what might otherwise have seemed arcane.

Lavoisier's chemical revolution had, as we know, been based upon book-keeping. Matter could not be destroyed, and therefore the weights on both sides of a chemical equation must balance, like the money in the columns of his tax accounts. Nannies used to say that babies could change their weight at will: we know from handling indignant and wriggling toddlers what they meant, but scales prove them wrong. Indeed, everyone had assumed the indestructibility, or conservation, of matter for a long time, but nobody had stated it so explicitly and put it to such important use. He also experimented on heat-changes in chemical reactions (most obvious in burning), but found the results inconclusive and difficult to reproduce. His chemistry remained a science of weights, applied algebra.

When Alessandro Volta discovered in 1799 that an electric current flows if dissimilar metals are immersed in water and connected together, he thought that their mere contact was responsible. The discovery rang around Europe like an alarm bell, but as others joined in the investigation, the feeling gained ground that one could not get something, electricity, for nothing: as King Lear had said, 'Nothing will come of nothing: speak again'.[5] Wollaston inferred that a chemical reaction, corroding one of the metals, must be responsible; and indeed putting acid rather than water as the medium increased the effect. Davy's work of 1806–7, establishing that electricity can be used for chemical analysis, confirmed his view that electricity and chemical affinity were manifestations of one power.

Davy had earlier speculated about the connection of electricity and light, inventing the arc-lamp. Light set off chemical reactions, both the photosynthesis by plants that Priestley investigated, and also the blackening of silver salts that got Tom Wedgwood investigating with Davy what became photography.[6] William Herschel the astronomer had discovered heat-rays beyond the red end of the spectrum; and Wollaston and Johann Ritter found that invisible rays beyond the violet end would set off chemical reactions. Heat, light and chemical affinity were somehow connected. It had long been clear that the force stored in a raised weight could be converted into motion; Romantic thinkers like Friedrich Schelling, associated with Ritter, taught in their Naturphilosophie that all force was one, though manifested in different forms. Armed with this notion, Oersted[7] established the inconvertibility of electricity and magnetism; and Faraday, with the first dynamo and electric motor, showed how electromagnetism could result from, or in, motion. In the USA, Joseph Henry (later first director of the Smithsonian Institution) built the biggest electromagnet of the day, raising great weights. Publics flocked to his demonstrations, as they did to lectures at the Royal Institution and elsewhere.

Amid this newly-created and considerable public interest, many men of science in the first half of the nineteenth century groped towards an understanding that force was somehow never lost any more than matter could be destroyed. But it was hard to measure force outside the narrow confines of mechanics. If we apply a metaphor from accountancy again, to make conservation (or correlation, as it was sometimes called) of force a properly-scientific principle, it must be quantified so that we can balance the equation of light and magnetism, for example. Euros, yens, dollars and pounds can be converted into each other; there are exchange rates. Thinking, measurements and calculations were required if such a thing were to be done for forces.

John Tyndall, in his popular *Fragments of Science* (for 'unscientific people'), sketched the two men generally held responsible for formalising the hunch of Davy, Faraday and Naturphilosophen.[8] Julius Robert Mayer and James Joule were awarded the Royal Society's Copley Medal: Joule first in 1870, and then Mayer, in a belated attempt to do him justice, in 1871. Mayer was a ship's doctor, who bled a sailor in Java in 1840 and noticed that the blood was bright red. Though it came from a vein, it looked like blood from an artery, full of oxygen. Back home, as an obscure physician in Heilbronn, he pondered this, and concluded that in a hot climate the body does not need to burn fuel to keep warm, and so uses less oxygen. He generalised over succeeding years, sending papers to Liebig for his journal, and came up by 1845 with a roundabout way of calculating from data on the expansion of gases how much mechanical work is needed to produce a definite quantity of heat. Nobody took any notice.[9] Mayer had a nervous breakdown. Later, his became a wonderful story of neglected genius.

Joule was a Manchester brewer interested in science, hence, like Lavoisier, used to keeping careful accounts of processes for tax purposes. A pupil of Dalton, he first investigated electric motors, and the heating power of electric currents. But measurements here were tricky and not generally convincing; so, deserting electrochemistry, he became famous with his experiments on heating water by stirring it.[10] He designed very accurate thermometers, and an insulated vessel in which water was churned by a paddle wheel driven by a weight falling through an exactly measured distance. He duly found that the work done by the weight falling produced a definite increase in heat: he could show directly what Mayer (unknown to him) had inferred. These experiments were reported in 1845, and the definitive paper on their improved version was read to the Royal Society in June 1849. Two years before that, in May 1847, he had delivered a public lecture at the Manchester Literary and Philosophical Society, and reported in the local newspaper, on 'Matter, Living Force and Heat' in which he had set out the general principle that force, like matter, was indestructible. But until his encounter with William Thomson at the Oxford meeting of the BAAS in June 1847, who ensured that his paper made a buzz, he remained marginal and unappreciated by a scientific community becoming increasingly professional. Happening on his honeymoon in Switzerland two weeks later to meet Thomson again, he induced him to collaborate in measuring the temperature at the bottom and top of waterfalls to see if there was an appreciable difference.[11] Later they worked together on more serious researches on heat and gases that led to the idea of an absolute zero of temperature, and ultimately to the liquefaction of air.[12]

Also in 1847, Hermann Helmholtz,[13] trained as a medical doctor, whose work was on the frontier of physiology and physics, published in Berlin an essay on conservation of forces. Because it was philosophical and methodological, rather than experimental, he had been unable to publish it in a scientific journal: by this time, Naturphilosophie was in bad odour, and positivism was in the air. But in 1852–3, Tyndall's translation was published in *Scientific Memoirs*, a journal dedicated to bringing foreign work before English-speaking scientists – an enterprise of Tyndall's and Huxley's that did not find a large public. In February 1854, conservation formed the topic of a professorial popular lecture Helmholtz delivered in Königsberg, 'On the Interaction of Natural Forces'. Again, Tyndall translated it; and, in due course, Helmholtz translated writings of Tyndall's into German. Like Liebig, he became well-known and much admired in Britain. It is curious that such important scientific works as Carnot's, Mayer's, Joule's and Helmholtz's should have been deliberately aimed at unscientific publics: all these authors were, at the time, in different ways outside the mainstream of physical science, though Helmholtz was to become a central figure, immensely respected, ennobled (like Thomson, who became Lord Kelvin) and reigning over his institute in Berlin.

Helmholtz's popular lectures gave him the chance to do the generalising and philosophising that the scientific community had found alarming. He was wonderfully wide-ranging, and capable of seeing the wood where many contemporaries saw only the trees. In the 1854 lecture, he began with reference to the 'new conquest of very general interest ... recently made by natural philosophy'.[14] He presented himself as a synthesiser, moving from trip-hammers to air guns, to steam engines and to electric motors. He concluded, stating clearly what others had more dimly perceived, that:

> Nature as a whole possesses a store of force which cannot in any way be either increased or diminished, and that therefore the quantity of force in Nature is just as eternal and unalterable as the quantity of matter. Expressed in this form, I have named the general law 'The Principle of the Conservation of Force'.

The term 'force' used for the German 'Kraft' had scientific connotations, because it came in the usual statement of Newton's laws of motion, as a translation of the Latin 'vis'; but it had always also been, and remained, in frequent and informal use. The word 'energy' was equally informal, but in 1807 it had been used in an exact scientific sense by Thomas Young in his lectures at the Royal Institution, discussing mechanics, and in 1852 Thomson took it up – whereupon it soon replaced 'force' in the context of physics.

The French Academy of Sciences had refused in the eighteenth century to examine any more proposals for perpetual-motion machines, and Helmholtz's principle confirmed that such things were indeed impossible. He referred also to the work of Carnot and Thomson establishing the law that heat flows spontaneously only from a hot to a cooler body; and went on to survey the cosmic implications of these laws, in the solar system, the growth of plants and animals, and the future of the Sun as its fuel is gradually consumed. He ended with a splendid peroration on their consequences for the human race:

> Thus the thread which was spun in darkness by those who sought a perpetual motion, has conducted us to a universal law of nature, which radiates

light into the distant nights of the beginning and of the end of the history of the universe. To our own race it permits a long but not an endless existence; it threatens it with a day of judgement, the dawn of which is still happily obscured. As each of us singly must endure the thought of his death, the race must endure the same. But above the forms of life gone by, the human race has higher moral problems before it, the bearer of which it is, and in the completion of which it fulfils its destiny.

As chemistry had had alchemists in its murky past, so classical physics in this vision had speculative inventors of impossible devices. Looking forward, Helmholtz had realised that conservation of energy implied that it must be possible to express all forms of energy in the same 'dimensions' of mass, length and time. Classical physics became the science dedicated to finding these conversion rates, and achieving standard units. Painstaking manipulation of precision equipment became the order of the day.

In Glasgow, Thomson combined mathematical prowess acquired in Cambridge[15] with exact equipment required to make electrical measurements associated with telegraphy. It was he who, after the first transatlantic cable had speedily burnt out, worked out how long-distance undersea cables should be worked. In Bonn and then Berlin, called there from distant Königsberg, Helmholtz applied the inventive talent which had enabled him to make the first ophthalmoscope (to look into eyes) to physical measurements. Back in Cambridge, Clerk Maxwell was in 1871 called to be the first director of the Cavendish Laboratory, founded with money from the Duke of Devonshire's family, and to preside over the first degree courses offered there in physics. He died young, and rather suddenly; his successor was not his sidekick and biographer, William Garnett[16] (who went to take charge of the College at Newcastle, and then of the newly-founded London polytechnics), but Lord Rayleigh, the son of an ennobled industrialist, whose forte was accurate measurement. An instrument-making company was set up in association with the laboratory, and the determination of units in competition with Berlin became the order of the day. This seemed to some at least rather like what Thomas Kuhn in the twentieth century described as 'normal science', filling in the big picture sketched by the pioneers of energy conservation, claiming the promised land seen from that mountain-top.[17]

The fruit of exact measurements backed by mathematical analysis sometimes attracted widespread public interest. Thus, in September 1846, J.F. Galle at the Berlin observatory had detected Neptune, the planet predicted by Urbain Le Verrier, who had deduced from wobbles in the orbit of William Herschel's planet, Uranus, that it was being attracted by another further out. This confirmation of Newtonian theory created great excitement – which increased when it emerged that John Couch Adams, a Cambridge mathematician, had made the same prediction in 1845 but had had trouble in getting busy astronomers to look for his supposed planet. When James Challis at Cambridge, whose major project was to do with comets, did start looking, he missed the significance of observations he had made in August. To the French, claims for Adams, who had not published his prediction, looked like typically Anglo-Saxon bad faith; and there was a famous caricature of Adams gazing through his telescope at Le Verrier's notebook.[18] In the event, the Royal Society gave the Copley Medal to Le Verrier in 1846, and two years later awarded it to Adams. Most publics could not follow the mathematics involved, but could appreciate the power of theory that led to such astonishing predictions.

François Arago, Permanent Secretary of the Paris Academy of Sciences, had suggested to Le Verrier an investigation of the orbit of Uranus. He was himself much interested in the nature of light; he devised a crucial experiment to settle once and for all the debate over whether light was a wave-motion, or particles. If it were a wave, it would go faster in air than in a denser medium such as water – while the opposite would be true if it were particles. Léon Foucault in 1850 undertook to perform the experiment, using an apparatus with rotating mirrors to make, in 1862, direct measurements of the velocity of light – and confirmed that (in accordance with what most by then expected, from other lines of reasoning) it was a wave motion. Here again, classical physics could deliver astonishing results; and Tyndall could take the wave theory as an excellent example of an established truth, a triumph of enlightened and enlightening science.

These complicated inferences were hard to follow for those outside the magic circle of physicists, but the conclusions could be appreciated. For the laws of thermodynamics, the conservation and degradation of energy, it was more difficult, but two professors of physics, Peter Guthrie Tait and Balfour Stewart, turned to natural theology in their initially anonymous *Unseen Universe* (1875) and its sequel, *Paradoxical Philosophy* (1878).[19] *The Unseen Universe* was a great success, reaching its sixth edition by 1876 – much faster than better-remembered works like the *Origin of Species*.[20] The authors' attempts to show, in opposition to Tyndall's Belfast Address, that energetics was fully compatible with orthodox belief made the books controversial, even to those like Maxwell who were sympathetic.[21] He remarked that the authors[22]

> Avail themselves of the general interest in theological dogmas to imbue their readers at unawares with the newest doctrines of science. There must be many who would never have heard of Carnot's reversible engine, if they had not been led through its cycle of operations while endeavouring to explore the Unseen Universe. No book containing so much thoroughly scientific matter would have passed through seven editions in so short a time without the allurement of more human interest.

Maxwell faced up to the problem, reviewing *Paradoxical Philosophy* in *Nature*, that there were limits to the legitimate application of scientific methods, and that science should be kept distinct from theology (or politics). And he concluded:

> The progress of science, therefore, so far as we have been able to follow it, has added nothing of importance to what has always been known about the

physical consequences of death, but has rather tended to deepen the distinction between the visible part, which perishes before our eyes, and that which we are ourselves, and to shew that this personality, with respect to its nature as well as to its destiny, lies quite beyond the range of science.

Clearly, the books had worked: thermodynamics had entered into the purview of the chattering classes.

Maxwell was himself a good populariser, but not in the same league for sales. His generally positive view of the books was at odds with that of Clifford, whose lengthy review was blistering even in its toned-down form in his posthumously collected *Lectures and Essays*[23] – it was originally in the *Fortnightly Review* rather than a scientific periodical. Clifford explained current ideas of the luminiferous æther, point-centre atoms, continuity, consciousness and the laws of thermodynamics, thus promoting public understanding himself, but denied that they had any bearing upon, let alone gave any support for, religious beliefs:[24]

> A sketch of the beliefs and yearnings of many different folk in regard to a life after death leads up to an attempt to find room for it within the limits of those physical doctrines of continuity and conservation of energy which are regarded as the established truths of science. In this attempt it is necessary to discuss the ultimate constitution of matter and its relation to the ether. When, by a singular inconsequence in writers possessing such power in their right minds of sound scientific reasoning, room has been found for a future life in the manner indicated above, it is discovered that there is room for a great deal more.

He ended with a peroration denouncing Christianity, 'which has made its red mark on history, and still lives to threaten mankind':

> You have stretched out your hands to save the dregs of the sifted sediment of a residuum. Take heed lest you have given soil and shelter to the seed of that awful plague which has destroyed two civilizations and but barely failed to slay such promise of good as is now struggling to live amongst men.

Clifford's trenchant views of scientific method were published posthumously, in his *Common Sense of the Exact Sciences* (1885).[25] Clearly, speculative treatments of contemporary physics could arise formidable passions, and there is no doubt that the *Unseen Universe* contributed heat and light to public understanding of science.

Physics, along with evolutionary biology, thus enjoyed a high profile; and some of its practitioners became well-known figures. But Thomson became a Lord (whereas Newton had been a knight, and Davy a baronet) more for his work on the telegraph than for his thermodynamics.[26] Science is an intellectual, a social and a practical activity; and classical physics developed close links with engineering, which by the 1870s could no longer be based upon rule of thumb.[27] Engineers had to know physics, and might, like William Rankine, have an important place in the science; and the physicist might, like Thomson, be an engineer. This was fairly new. Sir George Cornewall Lewis, a cabinet minister in mid-century, remarked that the physical sciences, when dealing in abstractions like perfectly elastic beams, were certain; but, in the real world, they were as uncertain as moral science.[28] But science and technology were not, and never are, quite separate. The public was astonished (despite accidents and uncertainties) not only by railways, speeding passengers, and telegraphs, sending messages at the speed of light; but also by great bridges and tunnels, and by enormous steam-engines, gleaming with polished steel and brass, for pumping water and for driving ships. Charles Parsons, son of the astronomer Earl of Rosse, further applied thermodynamics in building *Turbinia*, the tiny ship driven by a steam turbine that astonished spectators at a naval review by her speed and manoeuvrability.[29] In the USA, the West was opened up by the transcontinental railroads and the six-shooter; and, following the success of US exhibits (especially guns) at the Crystal Palace, Joseph Whitworth visited the USA to study industrial methods there, notably the beginning of mass production.[30] Whitworth reported on exhibitions he saw there, and his colleague George Wallis on education in art and design.

Back home, Whitworth introduced standardised nuts and bolts in Britain; and was deeply involved in the armaments business as rifles and carbines came to replace muskets, and artillery was modernised and made much more formidable with accurately bored rifled barrels and breech-loading. Alfred Krupp in Germany was one of his great rivals in big-gun making; another nearer home was Armstrong in Newcastle, whose breech-loading guns and armoured battleships supplied Japan with its formidable first modern navy. Those who suppose that the military–industrial complex is a twentieth-century phenomenon should note how much space in *Reviews*, and how many hours in lecture-programmes at the Royal Institution and elsewhere, were devoted to military matters.

War against Russia in 1854 in the Crimea and the Baltic (in alliance with France) had not gone as well as it should have, and there were correspondents like William Russell to send back reports of failures, incompetence and setbacks to newspapers.[31] Then there was in Britain great alarm about the Indian Mutiny of 1857–8, and real fear of French invasion in the 1860s. The USA was locked in civil war, Prussia was fighting Austria and then France, and Italy was being unified. Meanwhile, colonial wars were going on around the globe. Armaments involved chemistry too, and Frederick Abel's lectured on gun-cotton, cordite and other explosives. Alfred Nobel's development in Sweden of dynamite also aroused great interest, but classical physics as a science that brought results as well as arousing passions about materialism drew crowds. Even disasters, bridges that collapsed, ships that sank, factories that blew up, trains that crashed, could be seen in a positive light: they prompted investigation and experiment, leading to progress.

Such confidence could not last. Physics was to become exciting in a different way as the big picture that Helmholtz, Thomson, Tyndall and others had sketched turned out not to fit the world. In 1884, the BAAS migrated across the Atlantic, and met in Montreal. Afterwards, Thomson was invited to Baltimore, to the new Johns Hopkins University, where he conducted a seminar for American professors, on the luminiferous æther and the wave theory of light. A 'cyclostyled' version of the proceedings was circulated, but it was twenty years before Thomson wrote his work up, and by then things had moved on, partly as a result of those sessions.[32] They had been illustrated with experiments and models: Thomson was a hands-on person, who loved simple, everyday materials like cobblers' wax and string as well as the highly sensitive apparatus needed for electrical measurements. Because various crucial experiments had confirmed beyond reasonable doubt that light was a wave motion, there must be something to be waving. We are familiar with waves by the seaside: water is the medium there. The æther was invoked as the medium for light, and for infra-red, ultra-violet and other radiations. It must penetrate ordinary matter, and be present in the most perfect vacua. It was not directly detectable; and it was a very curious substance, putting up no resistance to planets and comets in their motions, and yet stiff enough to vibrate.[33] Among the participants, 'co-ordinates' as Thomson punningly called them, were Albert Michelson and Edward Morley, both well-known for precision measurements. They decided to see if the Earth's motion through the æther could be detected, using an interferometer, a device with which Michelson (who had come into science from the US navy) determined the speed of light, the exact length of a metre, and various astronomical distances. In 1907, he became the first American scientist to win a Nobel prize – that great twentieth-century source of public admiration, if not understanding.

In a boat, to go downstream with the current and back against it always takes longer than to cover the same distance in still water. So it should be with light. If the Earth is ploughing through the æther, then a beam of light sent forwards and reflected back will take longer on its journey than one sent out at right angles to the Earth's motion: they will thus 'interfere' with each other when brought together, and generate a pattern of bands. Michelson and Morley set up their sensitive apparatus in Cleveland, Ohio, and made measurements with it turned in various directions, at peaceful times in the middle of the night. Within the limits of experimental error, the result was always negative. One solution was to say that the Earth was carrying the æther with it; but that seemed rather odd. George FitzGerald, an Irish physicist who carried forward Maxwell's work on electromagnetic radiation, suggested in 1889 that motion through the æther might cause material objects to contract, so that the effect of motion would be imperceptible. He was not very interested in priority, and his idea was taken up and further developed by Hendrick Lorenz in the Netherlands. Clearly, the æther was very paradoxical stuff. This work did not mean the end of æther physics, which continued to flourish in Germany,[34] and in Cambridge with Joseph Larmor and others;[35] it was by no means sterile or boring, as Oliver Lodge's broadcasts of 1925 indicate.[36] Nevertheless, classical physics that had seemed so clear was beginning to look mystifying – paradoxical philosophy indeed.

Meanwhile, other forms of radiation were joining light: first radio waves found by Helmholtz's student Heinrich Hertz confirmed predictions in Maxwell's equations.[37] This was, at the dawn of the twentieth century, to show its practical value with the work of Guglielmo Marconi. Meanwhile, various other unexpected and perplexing forms of radiation came within the purview of science. Air is an insulator, but Faraday had found that electricity would pass through a glass tube when most of the air was sucked out. The tube glowed, in what was an ancestor of our fluorescent lights. As the pressure was further reduced, the glowing column broke up, and disappeared; but something streamed out in straight lines from the negative terminal, called the cathode, causing a greenish glow. As pumps improved, Faraday's admirer Crookes took up the investigation, and found that these 'cathode rays' cast shadows, made tiny paddle-wheels rotate, and were deflected by magnets.

He did spectacular demonstrations,[38] but could not provide any good explanation. On the Continent, the rays were generally believed to be a wave motion, akin to radio waves. When, in 1895, Wilhelm Konrad Röntgen in Wurzburg discovered X-rays, emitted from a cathode ray tube and causing fluorescence, that boosted the idea, because they certainly seemed to be waves too. Röntgen was an experienced and careful experimentalist, and investigated the rays systematically. The first X-ray photograph was of Frau Röntgen's hand; and the discovery, so very unexpected, caused wild enthusiasm: the medical implications were seen at once, and also the jokey possibilities of being able to see through clothes or walls. Understanding, scientific or public, was slower. In Adelaide, Australia, William Bragg and his son Lawrence saw the possibility of applying X-rays to reveal the structure of crystals, providing almost-direct evidence for atoms: Lawrence went on to direct the Cavendish Laboratory after Rutherford, and then the Royal Institution.[39]

Antoine Henri Becquerel found further rays in the following year, emanating from uranium, again causing fluorescence. The phenomena were confirmed by Kelvin and others. Marie Curie, with her husband Pierre, investigated what was happening, from a chemical perspective: she found it was characteristic also of thorium, called it radioactivity, and then isolated two new and more powerful sources, the previously unknown elements radium and polonium. Three different kinds of rays were identified as α, β and γ; the first were positively charged, the second negatively, and the third were analogous to X-rays. The rate of radioactive emission from a given element seemed to be constant, independent of chemical combination or physical circumstances: there did not seem to be any obvious cause. This was rather mystifying, because chemical reactions were sensitive to conditions.

J.J. Thomson, Rayleigh's successor at Cambridge (and no relation of Kelvin), resolved to test whether cathode rays were waves or particles with a crucial experiment. While not very skilled in manipulation, he had excellent vacuum pumps and well-trained assistants. If the rays were composed of negative particles, they should be deflected by an electrical field as well as by a magnetic one; but nobody had been able to observe that. The key was to get lower pressures than Crookes and others had attained; and Thomson managed that. He then very cleverly balanced the electrical deflection against the magnetic, so that the rays came back to where they had been focused without either. This meant that he could estimate the ratio of mass to charge of the particles. In 1897, in a lecture at the Royal Institution, and in a published paper, he inferred that they were sub-atomic, calling them 'corpuscles', and invoking William Prout's hypothesis that the elements were composed of simpler tiny particles.[40] Soon they were identified with the units of electricity invoked by Helmholtz and Johnstone Stoney, and called 'electrons'.

Meanwhile, in 1895, Ernest Rutherford had come from New Zealand to Cambridge on a scholarship funded by the profits from the Great Exhibition of 1851, to work with J.J. Thomson. In 1898, he was appointed Professor of Physics in Montreal. There, with the chemist Frederick Soddy, he investigated radioactivity, finding that uranium and thorium decayed into lead as they emitted the α, β, and γ rays. He called this the new alchemy: one element was actually changing into another. The prediction that lead found with uranium would have a different atomic weight from ordinary lead was confirmed by Theodore Richards at Harvard: the USA was by 1900 prominent for precision apparatus and its use. Soddy coined the term 'isotope' for atoms of the same element which had different weights: chemically, they all behaved the same. Most elements turned out to have two or more. Rutherford then came up with the idea that the atom consisted of a minute, central, positively-charged nucleus, with electrons in orbits around it.[41] This was completely anomalous, because the electrons should have very rapidly spiralled into the middle and crashed there with enormous emission of energy, and yet almost all atoms were very stable. Dalton's atoms had been billiard-balls, solid and indestructible, all those of an element weighing the same. That now seemed untrue.

London, Paris, Berlin and Cambridge (England or Massachusetts) were the sort of places, centres of excellence, where great scientific discoveries might be expected; but, just as Dmitri Mendeleev had classified the elements in St Petersburg, so Wurzburg, Montreal, Zurich and Adelaide also turned out to be where major scientific innovation happened. Mendeleev was lionised in London; Röntgen duly called to Munich; Rutherford to Manchester, and then to Cambridge to succeed J.J. Thomson; Bragg to Leeds, and then London; and Einstein, after correspondence with people in Cambridge,[42] to Berlin. Great centres can attract people from the periphery of the scientific world, but are not always where the excitement, the eureka moment, is to be found.

Kelvin, at the celebration of his fifty years as professor at Glasgow, had said that the word 'failure' summed up his efforts in physics, just before he signed up again there as a research student.[43] In 1900 he lectured at the Royal Institution, looking back at classical physics; but great man as he was, instead of dwelling on its triumphs, he depicted the clouds hanging over it.[44] There were two great big thundery ones. One was the question of the Earth's motion through the æther; while the second came through the dynamical theory of gases, in which

Maxwell had been a major player, in which they were seen as composed of molecules in motion. Some of their energy also went into vibration and rotation; and the assumption was that energy was equipartitioned; that is, equally divided between these modes of motion. This concept had been used to infer that argon, the gaseous element newly discovered by Rayleigh and Ramsay, had molecules composed of a single atom, unlike oxygen (O_2), nitrogen (N_2) and other familiar elements. This evidence came from heat measurements; but it created difficulties, and apparent contradictions.

Heat radiated from black bodies had meanwhile preoccupied Max Planck in Berlin, who from 1896 to 1901 sought a formula that would fit the facts over a wide range of temperatures, rather than those available which only worked for a part of the range. He found one; but only when he introduced a 'quantum of action' implying that energy came in chunks rather than continuously. He thus founded quantum mechanics, but never felt easy with this progeny, hoping for some better physical theory that would represent reality and not just, as he saw it, save the appearances. If light energy came in packets, it looked as if the particle theory, disproved in a number of crucial experiments, was, after all, more plausible than its wave, or undulatory, rival. Albert Einstein was to get his Nobel Prize for his work of 1905, demonstrating the power of quantum theory in accounting for photoelectricity, not so distant from those researches of Tom Wedgwood a century earlier on photochemistry; and Niels Bohr, coming to work with Rutherford, used quantum theory to explain atomic stability, and the spectra emitted from the atoms in hot gases. In these years around 1900, the either/or logic of crucial experiments, and ideas of causation, were put in doubt. The fruits of the new physics were wireless communications, X-rays (for Lawrence Bragg's injured arm, for example) and glowing radium figures on dials of clocks; but it was so exciting because it seemed a defiance of common sense, restoring wonder, raising the mind above the merely useful.

Walter Bagehot, the political economist, had in 1857 compared public attitudes to science in early nineteenth century with current more jaundiced ones:[45]

> The great discoveries in our knowledge of the material world were either just made, or were on the eve of being made. These enormous advances, which have been actually made in material civilisation, were half anticipated. There was a vague hope in science. The boundaries of the universe, it was hoped, would move. Active, ardent minds were drawn with extreme action to the study of new moving power; a smattering of science was immeasurably less common then than now, but it exercised a stronger dominion, and influenced a higher class of genius. It was new, and men were sanguine. In the present day, younger men are perhaps repelled into the opposite extreme. We live among the marvel of science, but we know how little they change us. The essentials of life are what they were. We go by the train, but are not improved at our journey's end. We have railways, and canals, and manufactures, – excellent things, no doubt, but they do not touch the soul. Somehow, they seem to make life more superficial.

He wrote just before the *Origin of Species* rekindled interest among the world-weary; later, fin-de-siecle gloom was to be lifted by the dramatic changes in physical science.

The lectures at the Royal Institution and similar places where these things were explained were not accessible to a very wide range of publics – the governing classes indeed, about whom Bagehot wrote. But in 1904 the British Association was due to meet in Cambridge, and the Prime Minister, Arthur Balfour, was chosen as President.[46] He was a Cambridge graduate, with a gentlemanly interest in and respect for science. He was educated chiefly at home by his mother Blanche, sister of Lord Salisbury who was to be his patron and predecessor as Prime Minister; and then at Trinity College, where he got to know Lord Rayleigh, and Henry Sidgwick, his tutor – the uneasily agnostic utilitarian, first layman to be Professor of moral philosophy at Cambridge, and founder of the Society for Psychical Research. Balfour became astonishingly well-connected: one of his sisters was married to Sidgwick, and became a great pioneer of women's education at Cambridge; another to Rayleigh; while Sidgwick's sister was married to Edward Benson, Archbishop of Canterbury. He could thus get good advice on anything to do with philosophy, physics or theology within the family. He had in 1879 published his first book, *A Defence of Philosophic Doubt*, where he boldly and wittily exposed science to the sort of sceptical doubts that many contemporaries were applying to religion. His conclusion was that:[47]

> Science is a system of belief which, for anything we can allege to the contrary, is wholly without proof. The inferences by which it is arrived at are erroneous; the premises on which it rests are unproved. It only remains to show that, considered as a general system of belief, it is incoherent.

He went on to do that, but he was by no means an opponent of science. Provided that it was recognised as provisional, not wholly coherent and empirical, he had no problems with it. A conservative and a religious believer, he admired Darwin, and had no difficulty with Darwinian evolution as a scientific theory, accounting for known facts of biology and geology. It was scientism and naturalism, the beliefs that scientific explanation was the only satisfactory kind, and that they entailed materialism, with which he quarrelled.

The book made little stir. But Balfour became an important public man; not just his uncle's poodle, but successful in his own right when from 1887 put in charge of Irish affairs, and from 1891 Leader of the Conservatives in the House of Commons, and spokesman there when his uncle became Prime Minister. In 1888, as a prominent figure in politics, he was elected a Fellow of the Royal Society. In 1895 he published his second book, *The Foundations of Belief* – in fact the subtitle of the first book. The message remained much the same, but this time it sold very well, and aroused great curiosity and interest. A society was launched to pursue the ideas in a series of seminars involving eminent clergy, scientists and lay people.[48] He wrote that our various beliefs, about God, nature

180 *Classical physics*

and beauty, for example, are not coherent, but we should not worry about that because we are not and cannot be omniscient. All our beliefs are approximations at best, and we must not impose dogma – which included naturalism, a gloomy and selfish creed:[49]

> Man will go down to the pit, and all his thoughts will perish. The uneasy consciousness, which in this obscure corner has for a brief space broken the contented silence of the universe, will be at rest. Matter will know itself no longer. 'Imperishable monuments' and 'immortal deeds', death itself, and love stronger than death, will be as though they had never been. Nor will anything that is be better or worse for all that the labour, genius, devotion, and suffering of man have striven through countless ages to effect.

In 1895 classical physics still seemed in good shape; but as it became incoherent, and its provisional nature evident, Balfour's analysis seemed increasingly plausible.

Balfour chose to address the British Association on the 'New Theory of Matter': 1905 was Einstein's *annus mirabilis*, so quantum mechanics was not yet part of the story. A brisk run through some history, with Cambridge especially in the focus, led him to reflect what a revolution there had been. The wave theory of light, and thus the æther, went back a century; and the history of the serious study of electricity (now fundamental to matter, following J.J. Thomson's work) was not much longer. Matter occupied space only metaphorically, as soldiers occupy territory – motion and charge were its features:[50]

> But if the dust beneath our feet be indeed composed of innumerable systems, whose elements are ever in the most rapid motion, yet retain through uncounted ages their equilibrium unshaken, we can hardly deny that the marvel we directly see are more worthy of our admiration than those which recent discoveries have enabled us to surmise.

Physics had restored the sense of wonder; and its progress had undermined the credo of Cambridge men, for whom:[51]

> A well-established theory is not a mere aid to the memory, but it professes to make us acquainted with the real processes of nature in producing observed phenomena.

Science was indeed concerned with the inner character of physical reality, and not just with appearances; but there was no reason to believe that the world was simple, and every reason to expect incoherence. Physics could not be the whole story, or the final one. The imposing of patterns, following hunches in the manner of Whewell, interrogating nature until the required answer is obtained, looked to Balfour plausible as a vision of science:

The plain message is disbelieved, and the investigating judge does not pause until a confession in harmony with his preconceived ideas has, if possible, been wrung from the reluctant evidence.

This echoes Bacon's 'putting nature to the question', with its overtones of torture: it is like Davy proving that water free from air yields only oxygen and hydrogen when an electric current passes through, though appearances suggested that acid and alkali were somehow generated. For Balfour, if the new physics was true, then what he had learned as a student, and everybody had learned down to very recently, was wrong, and they had lived in a world of illusions about what they saw and handled.

This picture of science as provisional was very different from Tyndall's of forty years before, arrogant in the bold confident morning of classical physics. But it went down well with the scientific community; and, in 1920, Balfour, now an elder statesman, was asked by J.J. Thomson on behalf of the Council of the Royal Society if he would be President. He refused when he found that the President was not a figurehead; it is astonishing that after a century of distinguished researchers, the Society should have fixed upon him, and indicates the uncertainties of the post-war world and the belief that someone from the very centre of the 'establishment' would be ideal. He did however become, in various cabinets in the 1920s, in effect the minister for science; and when he died, Lord Rayleigh (son of the great physicist) wrote his obituary for the Royal Society. He remarked on Balfour's BAAS Presidency:[52]

> He delivered an address in which what were then novel views of the electronic structure of matter were discussed in their philosophical aspect. This was perhaps the first occasion when emphasis was laid before a popular audience on the glaring discrepancy between the new ideas of the atom, with its relatively vast inter-electronic spaces, and the old philosophic distinction which made shape a 'primary' property of matter, existing independent of the observer, while secondary qualities, such as colour, were thought to have no such independence.

Physics did not lose its prestige: Rutherford after all is supposed to have said that all science is either physics or stamp-collecting. But no doubt its provisional character was increasingly recognised, thanks to Balfour – it is not often the case that a Prime Minister takes on the role of a populariser of science. We must now turn to those who made a living out of it; and conclude by asking how successful was the whole project of making science accessible and popular, and converting people to scientific ways of thinking.

13 Promoters and popularisers

Few Prime Ministers or Presidents in the Anglo-Saxon world have had the qualifications or connections to be plausible leaders of the scientific community, and effective popularisers of science. In Britain and the USA, however, there were a few prominent scientists who played important political roles, thus directly or indirectly affecting the popular understanding of science. Lyon Playfair, promoting the Great Exhibition and Benjamin Franklin, and then Alexander Dallas Bache superintending the US Coast Survey, would be examples.[1] The chemist Sir Henry Roscoe entered Parliament (representing Manchester), where he championed the metric system and the opening of museums on Sundays – but with limited success.[2] Darwin's friend and neighbour Sir John Lubbock (later Lord Avebury) was a prominent Member of Parliament, famous for promoting a public holiday in August, 'St Lubbock's Day', and also an original, accessible and very successful writer on insects (he kept a tame wasp), and a powerful figure in scientific institutions.[3] Hooker however deplored this wasting of talents in politics.[4] Huxley and others sat on Royal and Parliamentary Commissions, advising governments; in the small intellectual and political world, informal contacts were important in getting scientific ideas and information through to legislators. The Athenaeum Club, with Davy and Faraday among its founders, was an important meeting place for intellectual members of the establishment. In France things were rather different. From Napoleon's meritocratic Empire onwards through the nineteenth century, eminent men of science, including Laplace, Arago and Marcellin Berthelot, did occupy important places in government, having actual direction of affairs.[5] Similarly in Russia, administrative posts were often more attractive than those in universities to men returning from study abroad.[6] How far such people were lost to science, or were vital in promoting it, is an open question. If it is true that science is a game for the young,[7] then for distinguished professors to be shunted into something else may be a very good thing all round; and middle-aged scientists, as well as directing laboratories and research schools and sitting on committees, could also be the best popularisers at a variety of levels.

Nevertheless, by the 1920s, popularising had come to seem a soft option for scientists, whose reputation should depend upon original research. Thus J. Arthur Thomson, whose well-paid and successful popular writing also promoted natural theology, met the disapprobation of his peers.[8] But in the previous century it had been different – and perhaps again now, amid science wars and the desertion of science by the young, it will be different again. Davy, and his contemporaries in France, attracted great audiences and boosted their reputations thereby; and, later in the century, Agassiz at Harvard, Faraday, Owen, Huxley and Tyndall in London, Liebig in Giessen, and Haeckel in Jena did the same. It was as science became more professional, Huxley's 'Church Scientific',[9] that descending into the popular arena seemed beneath the dignity of serious practitioners: an attitude not yet extinct. Herschel had written accessibly about physics, and Maxwell wrote for the classic ninth edition of *Encyclopaedia Britannica* essays on 'atom' and 'attraction' that were reprinted in his *Scientific Papers* along with his highly-mathematical writings – there was no cause for shame in being a good communicator, making vivid use of metaphor and rhetoric.[10]

The style of scientific communication had however begun to change by mid-century towards the compressed writing, full of passive verbs, abstract nouns and jargon that nowadays schoolchildren taking science have to learn. Symbols, equations, tables and graphs were further deterrents to the ordinary educated but unspecialised readers dipping into the Royal Society's *Philosophical Transactions* or *Proceedings*, where in Davy's time they would have found much to interest them. Publications of more specialised institutions like the Chemical Society of London were even more forbidding.

There were two main reasons. First, as the scientific community grew, so did the amount of research, and hence the number of papers offered; and editors had to demand compression. Background and implications could no longer be dilated upon in a leisurely manner. The amateur reader's attention need not be sought: professionals would scan the abstract at the head of the paper, or separately published. More papers meant shorter papers, and even so, journals became fatter. Second, the same growth went with increasing specialisation, knowing more and more about less and less, and with educational systems in which scientists were increasingly separated from those studying languages or humanities. There were enough people who spoke or read the languages of the sciences to make publications in them worthwhile. And then there is a certain satisfaction in acquiring and using a learned tongue, a private language inaccessible to ordinary people, as we all well know; scientists have sometimes been happy for their readers to be blinded with science, rather than enlightened. Thus doctors in our enlightened days do not always find it easy to maintain patience with the too-well-informed patient.

These things meant that popularisers became more necessary. There were different publics, who wanted to understand at levels appropriate for them the science which was going to change their lives, and for which they were going to pay. Mary Somerville had managed, as specialisation got underway, to get science across to the learned world.[11] She was much admired by her male contemporaries, and a sculptured bust was placed in the Royal Society, from which she was debarred. She had great talent, particularly in mathematics, but saw herself as unoriginal; in so far as that was true, it meant that the accounts she gave, and the implications she drew, were mainstream and highly acceptable to contemporary men of science. One of them, Whewell, published his word 'scientist' first in a review of her work. Where she differed from someone like Davy's able but unoriginal successor at the Royal Institution, William Thomas Brande,[12] was that, as a woman, it was almost impossible for her to give lectures, or to read papers about her work at a learned society. She wrote accessibly, but not very differently from how Fellows of the Royal Society, like her husband, would do.

We have met other women, improving public understanding by translating scientific writings, notably of Humboldt, Oersted and Comte. This was not an activity confined to women, for Oken and Helmholtz were translated by men; but translation was an important activity, which could have major implications for public esteem. Liebig for example seems to have had a higher profile in Britain, for his *Familiar Letters* and work on fertilisers, than in his native country. J.F.W. Johnston, conversely, became very well-known in Germany and Scandinavia through translations of his *Chemistry of Common Life* and other writings.[13] Textbooks were another kind of popularisation where the author would, as a rule, present the received view, rather than highly original interpretations. This became increasingly the case as formal instruction in the sciences grew. Parkes' *Chemical Catechism* and Jane Marcet's *Conversations on Chemistry* (and its successors and imitators) were informal introductory textbooks, written before there were syllabuses to get through. But even when there were exams to pass, textbooks might differ considerably in how they were organised, and what they presented as most important.

Thus at University College, London, students got a factual and analytical course of chemistry from Edward Turner, while those at King's College in the same overarching University of London got physical chemistry from an admirer of Faraday, J.F. Daniell, who declared in his preface that he aimed:[14]

> To present to students of chemistry an elementary view of the discoveries of Dr Faraday in Electrical Science. From the very first publication of his *Experimental Researches in Electricity*, I have felt that from them Chemical Philosophy will date one of its most splendid epochs; and perceiving, at the same time, that the results bear upon them the great impress of natural truths, namely, that they simplify while they extend our views, I have, from the first, availed myself of them in my instruction to my classes.

The textbook which can simplify and extend is doing very well; and the great example in nineteenth-century chemistry was Dmitri Mendeleev's. Linnaeus and Lavoisier had simplified the technical languages of natural history and of chemistry, so that all adepts would know precisely to what the name of a plant or a compound referred. Linnaeus had also adopted a taxonomic system for botany, the sexual system: for chemistry it was more complicated, and textbooks by the 1850s were fat and rather shapeless – there were a huge number of little-related facts to master. With his Periodic Table, organising the elements into natural groups based upon atomic weights and chemical properties, Mendeleev had a key to presenting chemistry as a coherent discipline, in which predictions might be made: the 'chemical philosophy' for which Daniell had sought, realised at last. This did help the understanding of that public composed of the many students who, by the 1870s, were having to learn chemistry professionally.

Science may be an imaginative activity, but the imagination cannot flow free. In poetry, it is subject to rules of language and form, in sculpture and painting to the nature of materials; in science, by facts. Just as rules can be broken by the maestro, and materials made to do breathtakingly different things, so (as the origin of the word reminds us) 'facts' are made as much as discovered – different phenomena become interesting, relevant and reproducible as science moves on. But there is a great deal of established knowledge that has to be learned. The profession of science, like that of law or accountancy, involves a great deal of grind before it starts to become fun. So, while elementary textbooks, in the tradition of Parkes and Marcet, can set out to be enthralling, to draw the reader in, this cannot last. As everyone knows from dull schooldays, working through a syllabus is a good way of getting inoculated against a subject. Writing more advanced textbooks is an art, involving order, compression and clarity, but its products are inevitably going to be unattractive sources of public understanding to those who do not need to qualify formally.

What they want is the big ideas and the fascinating examples without all the tedious detail; and this professional popularisers supplied throughout the nineteenth century. There were many books, often small in format to fit into a pocket, in which writers set out to present the latest discoveries, catching and holding the attention of their readers in various ways. Many were in effect hack writing, by authors like Jeremiah Joyce.[15] Joseph Guy, who had taught at the Royal Military College at Great Marlow, wrote *Elements of Astronomy* for the same publisher as Joyce: Baldwin, Craddock and Joy. They called their series 'popular school books'. Guy's contains examination questions at the back, and is priced at five shillings, with discounts for schools.[16] It is surprising that, at this date, there should have been sufficient schools teaching elementary science to make this enterprise worthwhile, but it was said to be as appropriate for 'private students' as for 'public seminaries' (schools) and may, like Jane Marcet's *Conversations*, have been used by mothers, governesses and tutors in enlightened middle-class homes. Like Parkes' *Catechism*, the book makes use of two sizes of type; larger for the main story, smaller for the details, important in a world where some children, like Mill, were crammed early:

> That work must surely possess some advantages, that can be perused by the younger scholar without perplexity, and by the more advanced student without deficiency. ... Everyone is aware of the impropriety of surcharging the bodily organs, – but overloading the yet unexpanded faculties of the mind, by the attempt to fill it with a too great redundancy of ideas in a first course, is equally fruitless and injurious.

The book, not surprisingly, aims at arousing wonder, but without dazzling or bewildering, and is in the tradition of natural theology:[17]

> In short, the compiler has been desirous not only to smooth the rugged avenues to knowledge, but to unlock the reluctant doors of the vestibule of astronomical science, and present to the youthful view URANIA, presiding amidst her spheres; not however in all the splendor of unveiled brightness, but with rays moderately attempered, that the mental eye of the juvenile intellect might be able, steadily, and undazzled, to contemplate something at least, of the harmony of our solar system and glories of the universe.

The reader was expected to detect in the heavens the work of an Almighty hand, looking in Newtonian vein through Nature up to Nature's God. Guy supported William Herschel's idea that the Sun was not a globe of fire, but an opaque inhabitable world, differing little from the planets except in having a luminous atmosphere. A previous owner of my copy has made notes at the back on that topic: not all science, popular or learned, has stood the test of time.

From the same publisher (and the catalogues at the backs of these books are fascinating) came Joseph Taylor's *Anecdotes of Remarkable Insects*,[18] with a splendid frontispiece showing Chinese travellers startled to encounter a very large praying mantis – a delightful mixture of Chinoiserie and natural history. This did not set out to be systematic – it is in the tradition of Pliny's *Natural History* rather than of the more austere writings of eighteenth- and nineteenth-century naturalists.[19] It contains poems by Anna Laetitia Barbauld, Mary Robinson,[20] William Cowper and others, many anonymous, on insects: their beauty, short lives, busyness, usefulness, social organisation and instincts are dilated upon, and squashing them deplored. This is thus an agreeable repository of tradition, lore, up-to-date science and literary allusion, making it a very sugared pill. It was published at a time when the serious study of entomology was beginning in Britain with the genial volumes of William Kirby and William Spence's attractive *Introduction*.[21] Insects had seemed rather ridiculous things to be interested in; but some, like bees, were important for making honey and fertilising flowers, and others like the brown-tail moth were pests worthy of a monograph.[22] In 1885 Vincent Holt published *Why not Eat Insects?*, an answer to agricultural depression as well as a culinary work.[23] He asked why we should be appalled at the thought of dining on caterpillars when we happily guzzle oysters and lobsters; noting Moses' and John the Baptist's taste for locusts, and that the Chinese found the unwrapped chrysalises of silkworms delicious when fried with egg-yolks in butter or lard, and seasoned with pepper, salt and vinegar. The message, conveying quite a lot of natural history, was that if snails (Holt used the term 'insect' loosely), cockchafers and other pests of garden and farm were carefully collected and eaten, they would furnish protein and the crops would be better.

Poisonous plants were the topic of another very necessary little book of natural history, illustrated with coloured plates, by Anne Pratt; published, in this case, by the Society for Promoting Christian Knowledge.[24] Her aim was threefold: to prevent the use of poisonous herbs in cookery, to caution those in the habit of nibbling unknown plants and to warn those who concoct herbal remedies – she could hardly have written that the book would be useful to the assassin, which it might be. She excluded the fungi, but even so there is a formidable range of poisons among the common plants of Britain and western Europe to be described. The daffodil even features among the 'suspicious' plants, its 'powerful and unpleasant odour' a give-away, producing headaches – while if eaten it will induce vomiting. So does the bluebell, and we are told that the slimy juice from its bulb was used stiffen ruffs in Queen Elizabeth's time. Folklore and natural history are happily compounded in this little book; although the focus prevents any systematic route into botany, it undoubtedly contributed to public interest in and understanding of botany, in the long tradition of Gerard's *Herball* and other works of the seventeenth century.

Also adorned with handsome coloured plates in small format was the parson–naturalist John George Wood's *Common Objects of the Microscope*, illustrated by Tuffen West.[25] His clear style and excellent illustrations made Wood one of the most successful popularisers of natural history in Victorian Britain; and he also took to the lecture circuit, visiting the USA twice. In larger format, his *Homes Without Hands*, dealing in a lively way with nests, dens and burrows, was another great success. He was no mere hack; he had good contacts with important naturalists, including Bates, Wallace and Waterton, and his books were reliable as well as enjoyable. His was serious popularisation; and so was that of his contemporary, Francis Orpen Morris, another parson–naturalist, the 'Gilbert White of the North', whose books on birds and on butterflies were particularly well-known. At his death, the eighth edition of his handsome butterfly book, with hand-coloured illustrations, in quarto format, was just on the point of appearing – it included a brief obituary, and reprinted the preface from the first edition of 1853:[26]

> An instinctive general love of nature, that is, in other words, of the works of God, has been implanted by Him, the Great Architect of the universe – the Great Parent of all – in the mind of every man. There is no one, whether old or young, or of whatever circumstances or rank in life, who can look without any feeling or emotion on the handiworks of Creation which surround him – who can behold a rich sunset, a storm, the sea, a tree, a mountain, a river, a rainbow, a flower, without some degree of admiration, and some measure of thought. He may, indeed ... be engrossed by some worldly care ... but ... whenever the mind is relieved from that overpowering feeling, the spontaneous thoughts which originate in the love of nature, will be sure to arise in his soul.

Entomology was particularly delightful. Such natural histories had a very wide appeal particularly to the nostalgia of the first generations to live mostly in towns and cities. At the end of the century, W.H. (William) Hudson, nostalgic as we know for the pampas of Argentina where he had grown up, wrote his *Birds in London* as a pioneer work of urban natural history.[27]

Coloured plates in unpretentious books were a feature of the mid-nineteenth century on, although delicate colouring as in Morris' books was still done by hand. Some may prefer the detail in black and white wood-engravings to the gaudiness of early colour printing. In the anonymous little book on timber trees, published by the indefatigable publisher of popular works, Charles Knight,[28] in his 'Library of Entertaining Knowledge', we find beautifully-executed cuts of trees, leaves and fruits, with an attractively readable text.[29] Here, delight and wonder were complemented by usefulness, as is proper in science; and there are good descriptions of industrial processes involving wood. We may be impressed by the sheer volume of information contained in this little volume, and the others in its series: the 'entertaining' aspect did not mean that knowledge was oversimplified, and our ancestors with their gas lights and oil lamps read hard and were prepared to concentrate. Acquiring some science was serious, though successful authors made it palatable to the public with their literary skills.

Contemporary with this little book was a Scottish work, *Popular Philosophy*.[30] This was 'upon Christian principles', and included moral reflections 'to excite devotional feelings in the breasts of the young'. In two volumes, it was an enlarged version of what had already appeared in a periodical. There are two engraved frontispieces: a day scene, and a night scene, both of which have lengthy commentaries. The first shows a romantic figure gazing into the picture, where there are sublime mountains (including volcanoes in eruption), a steam ship, a suspension bridge and a balloon hovering over all. There is also a whale, horses, birds and animals in a scene both sublime and picturesque. The other shows the Arctic winter twilight, with icebergs and Northern Lights, a comet, a sailing ship, kayaks, a dog sledge, Eskimos returning from a successful seal hunt to their igloo, a walrus unaware that a polar bear is about to attack it, a reindeer and a geyser.

These were features of up-to-date natural history (animal, vegetable and mineral), for polar voyages were the 'big science' of the day, and transport and communications were being revolutionised. The tone throughout is excited, whether over the richness and variety of things, the wonders of industry, or the ways of life of distant peoples. It starts in the Museum in Edinburgh, and then proceeds beyond, in order to demonstrate the wonder and harmony of the creation:

> The result of all our inquiries, is – that every page of this huge volume is emblazoned deep with gigantic characters, – every paragraph is illuminated by rays of celestial wisdom, – every line conveys a lesson of transcendent instruction, written not by human hands, but by the finger of Omnipotence itself, – and all things speak of a God!

Public understanding of science in 1826 thus involved awe and wonder, both about the variety and extent of the creation, and also about the possibilities of technology.

By the time Morris was writing, Huxley and others were seeking to establish physiology in Britain, where medical science was lagging behind what was happening in Germany and France. Anatomy had earlier been dogged by the need for human corpses, and Burke and Hare, who literally murdered to dissect (or to let Robert Knox[31] and his pupils dissect), made the murky world of grave robbing (so-called, perhaps because corpses had no human owner) very obvious and disgusting. When workhouses made unclaimed bodies available, grave robbing became unnecessary. But the exciting medical science was not to do with dead, but living bodies. Charles Bell, the Bridgewater author, had from dead specimens alone concluded that sensory and motor nerve endings were distinct; but the 'murderous' Francois Magendie, his contemporary, had used vivisection of cats to get to such understanding quicker, and take it further. To do physiology properly seemed to require such experiments, to the horror of many in Britain. Morris, whose work had emphasised animal sagacity and could not doubt that they felt pain as we do, was one of the opponents, writing in 1890 about the cowardly cruelty of those who experiment on living animals.[32] Earlier, in 1867, he had presented a petition at the House of Commons against animal experiments unless the gain to medicine was clear and certain. Charles Darwin was invited to give evidence to MPs, who 'treated him like a duke'. He accepted that some experiments had probably been tried too often, or done without anaesthetic unnecessarily, but believed that improvement of humanitarian feelings was the way forward – writing to his daughter in January 1875:[33]

> If stringent laws are passed, and this is likely, seeing how unscientific the House of Commons is, and that the gentlemen of England are humane, as long as their sports are not considered, which entail a hundred or thousand-fold more suffering then the experiments of physiologists – if such laws are passed, the result will assuredly be that physiology, which has until within the last few years been at a standstill in England, will languish or quite cease. It will then be carried on solely on the continent.

We are familiar with rather similar debates about cloning and stem-cell research, and about hunting with dogs.

The progress of science had thus, by the 1860s, begun to generate heretics, opposed to the direction that the leadership wanted. If all publicity is good publicity, they might count as popularisers – but some at least set out to be unpopularisers, just as William Morris and others were coming out against industry and its devaluing of craft skills and traditions. Science, pure or applied, was not simply wonderful and benevolent, indeed god-like, but to some publics at least might be devilish. Probably the leading figure in the agitation against vivisection was Frances Power Cobbe, from Ireland, born into a family including Anglican bishops and archbishops, and very well educated. She became a feminist, very suspicious of organised religion, but adhering to an undogmatic spirituality. She devoted her life to many good causes involving the position of women. In 1875 she was a founder, and for many years Secretary, of the National Anti-Vivisection

Society; but in 1898 she left it and formed a more thorough-going group, the British Union for the Abolition of Vivisection – in which she was prominent up to her death in 1904. The Society's agitation was so successful that stringent legislation was indeed proposed; and we cannot doubt, with Darwin, that some researchers were cavalier or callous in their attitude to animal suffering, doing poorly-planned experiments without anaesthetics. Huxley and his associates believed vivisection essential; but scientists came to accept that there would have to be control, if such research was not to be outlawed. In 1876, after much lobbying on both sides, the British Parliament passed the Cruelty to Animals Act, a weaker version of what Cobbe and her associates had wanted, but the first such law.[34] Other countries followed in due course. Public understanding was suddenly different: science was on the defensive, its morality questioned.

We are familiar with that continuing state of affairs. In medicine, there has been a long tradition of heresy or quackery, with patients suspicious of orthodox practitioners. In the nineteenth century, hydropathy (water cures) and homeopathy, with its doctrine of treating diseases with minute quantities of substances that would produce similar symptoms, were very popular. Charles Darwin tried various such 'alternative' therapies in his search for health; but his eminent doctor father was scathing.[35] He:

> observes that as long as he can remember there has always been something wonderful, more or less of the same kind, going on, and there have always been people weak enough to believe, and he says, slapping both knees, he supposes it will always be so.

He was duly unsympathetic about his son's 'dreadful numbness in my finger ends'. But, by the 1870s, Robert Koch and Louis Pasteur, using animal studies, were devising new vaccinations and treatments for terrible diseases; and producing statistical evidence as well as dramatic demonstrations to show that they worked. Roscoe noted that the death-rate from rabies used to be 15 per cent, but that Pasteur's treatment reduced that figure to less than 1 per cent; and he saw Pasteur's work as one of the great justifications of science in general, and of physiological research in particular.[36] Pasteur was himself a showman of some talent, who relished controversy; this was not an activity confined to quacks.[37] The pressure on animal experimenters to reduce, refine and replace, the 'three Rs' of twentieth-century best practice, began in the 1870s; and the 1876 Act was, with 'alkali inspectors' supervising the chemical industry, a beginning of legislative control over science and technology, perceived as a perhaps amoral and dangerous activity rather than simply philanthropic.

Vaccination against smallpox, literally using material from infected cows, was begun by Edward Jenner at the end of the eighteenth century, and despite early opposition and caricaturing, the procedure caught on. In 1853 it was made compulsory; and this outraged defenders of civil liberties, who found opposition easier now that smallpox was no longer the scourge it had been. Among the opponents of compulsion was Wallace, who indeed came to believe that vaccination was ineffective, and might indeed be dangerous:[38]

> The claims for vaccination were enormously exaggerated, if not altogether fallacious. I also now learnt for the first time that vaccination itself sometimes produced a disease, which was often injurious to health and sometimes fatal to life, and I also found to my astonishment that even Herbert Spencer had long ago pointed out that the first compulsory Vaccination Act had led to an increase in small-pox.

He declined membership of a Royal Commission to investigate; but, in 1898, brought out a pamphlet, circulated to every MP by the National Anti-Vaccination League, with the wonderful title *Vaccination a Delusion: its Penal Enforcement a Crime, proved by the Official Evidence in the Reports to the Royal Commission*. He hoped that in the future this would be seen as one of his most important and truly scientific works. Public suspicion of jabs and shots, fuelled by those whom the scientific establishment would see as heretics or cranks, is nothing new but also not extinct, as the controversies over MMR and Gulf War Syndrome indicates.

Wallace's[39] and Darwin's theory of evolution had, twenty years after their joint paper of 1858, itself become orthodoxy, as the distinguished opponents like Owen, Agassiz, Mivart, Gosse and Phillips aged or died, and younger scientists more or less accepted it as their guide through the perplexities of biology. Liberal-minded churchmen and other intellectuals also came to see development everywhere. Nevertheless, there were resolute opponents whose utterances and publications kept this also in the public eye as a controversial business. Thomas Cooper's little work, *Evolution, the Stone Book and the Mosaic Record of Creation*, from a respectable London publisher, appeared in 1878.[40] My copy has inside it a printed letter of Saturday 11 May 1878 asking the recipient, R. Ellis Esq., to be so kind as to assure the author that he received the book (to the address Thomas Cooper, Lecturer on Christianity, Lincoln – the Post Office must have been rather helpful, if that was sufficient). Cooper's life was extraordinary: apprenticed to a shoemaker, he became a schoolteacher and Methodist preacher, then a journalist, and an enthusiastic Chartist. Imprisoned for inflammatory activity, he wrote an epic libertarian poem, 'The Purgatory of Suicides'. He made contact with Disraeli and with Douglas Jerrold in his efforts to get it published. After it was, he lectured to radical groups, but was reconverted to Christianity and lectured on that instead. His book, an extended attack upon Darwinian evolution from the point of view of a literal believer in the Bible, shows evidence of wide reading in the sciences. Like other sermons against particular authors and doctrines, it no doubt got some of its readers going curiously to the forbidden fruit.

Heretics, cranks and quacks no doubt assisted publics in their understanding of science, because they demonstrated that it was not monolithic, fully systematic, unchanging and utterly reliable. Scientists always hoped that there was a gulf fixed between fact and fiction, and that science was firmly on the factual side. Nevertheless, ever since the days of Galileo and Newton and the 'Scientific Revolution', there had been works which could be labelled 'science fiction'. Some in this category were dialogues, where science was got across in imaginary conversations; others were adventure stories with a basis in science, actual or imagined. Galileo was condemned by the Inquisition for his dialogue, set in Venice with three characters discussing the counter-intuitive (indeed probably heretical) notion that the Earth goes round the Sun – all the more persuasive for its literary quality.[41] More happily, Bernard Fontenelle, permanent secretary of the Paris Academy of Sciences, published in 1686 a dialogue between an astronomer and a witty aristocratic lady on the same topic, coming down on Galileo's side.[42] Meanwhile, Robert Boyle's *Sceptical Chymist* had presented the arguments for an atomic, or 'corpuscular', world-view, especially in chemistry; his writing is verbose, and his characters wooden, so that this is a book more referred to than read.[43] Later, the dialogue form was used by Davy in his last works; by Benjamin Brodie, later President of the Royal Society, in his *Psychological Inquiries*;[44] and in the *Unseen Universe*. Brodie was dealing with the tricky frontier between bodies and minds, and dialogue gave him a vehicle in which various views, and anecdotes, could be conveyed to the reader. Science often does proceed dialectically, through arguments (perhaps conducted in a single head), and its conclusions are provisional: so the fictional form of the dialogue can work very well to get across controversial views, as philosophers since Plato have realised.

Other kinds of science fiction start in modern times with Johannes Kepler and his *Dream*.[45] His hero was transported (by witchcraft) during an eclipse to the Moon, where he found that its inhabitants took it for granted that their globe was the centre of the universe, and that all the other celestial bodies revolved about it: this lunatic notion indicated the equal absurdity of supposing that the Earth was the centre of things. Cyrano de Bergerac, himself chiefly remembered as a semi-fictional character with a formidable nose, wrote an account of travels to the Sun and Moon, where entertaining adventures awaited his hero in what was not only science fiction, but also utopian literature.[46] But the nineteenth century, as science and technology got serious, saw the appearance of one of the most resonant of all works of science fiction, *Frankenstein*[47] (1818). Written by the teenage Mary Shelley after an evening's telling of ghost stories by Byron and Percy Shelley on the shores of Lake Geneva, it has been analysed from feminist, racist and other angles, but from our point of view it is the powerful image of the scientist as sorcerer's apprentice that is most distinctive, and which (especially since the book was filmed) has been most prominent.

Young Frankenstein as a student was bored by the pedantic physics professor, but excited by the sweet-voiced Professor Waldman (based on Davy) lecturing on chemistry and the unlimited powers latent in science. He had been prepared for this by reading Agrippa, Paracelsus and others in the alchemical tradition. He worked hard on his own, collecting body parts from charnel houses, to create a living creature, somehow using electricity to galvanise it into action; but when his hideous monster came to life, he fled. It was innocent, but corrupted by having to make its way in a hostile world. It pursued Frankenstein's family and friends murderously, to induce him to create a mate. He began, but then, thinking how terrible it would be to start a generation of monsters, abandoned this second creation. The infuriated monster, having already encountered him on the sublime glacier, the Mer de Glace on Mont Blanc, pursued him into the Arctic. There the exhausted Frankenstein met an iced-in British ship, under the command of Robert Walton, determined against all odds to get to the North Pole: 1818 was just the epoch of those great polar voyages. After telling Walton his story, Frankenstein died; the monster intent upon suicide carried off the corpse of his maker; and Walton, a sadder and a wiser man, headed for home with his crew as the ice providentially opened up, his megalomaniac ambition subdued, and his thoughtless, fatal plan abandoned.

God had made Adam beautiful (in His own image); when he was lonely, He had made Eve as his companion, placed them in an earthly paradise, and continued to think in a fatherly way about them even after they ate the apple. Frankenstein's act of creation was the antithesis of all this. He had not thought through the consequences of his solitary and unnatural actions, trying to bypass the role of women in making life. He was incapable of unselfish love in his pursuit of glory and the intellectual pleasures of science devoid of ethical concern. No wonder that, in the twentieth century, the book seemed prophetic, so that applying the term 'Frankenstein vegetables' to genetically modified food, or using his name for nuclear energy, makes a powerful point instantly. Mary Shelley's book, seen mostly as just a horrific exercise in gothick fiction when published, has come to play a powerful role in how later publics have come to understand science: as an alarmingly powerful force, that must be overseen by ethics committees and legislators, much too much like the secret and forbidden knowledge of alchemists and magicians.

Much more positive in his attitude to science and technology was the French author Jules Verne, trained in law, who wrote opera librettos without any great successes. He then began writing novels, which were immensely popular tales of adventure and derring-do, involving extrapolated science. They were rapidly translated from French into many other languages. In 1864 came his *Journey to the Centre of the Earth*, followed soon after by *From the Earth to the Moon*, in 1870 by *Twenty Thousand Leagues Under the Sea*, and in 1873 by *Round the World in Eighty Days*. The detail, the human interest, the wise coyness (like Mary Shelley's) about how some of the science worked, all made the stories enthralling; and they have subsequently been made into very successful films. For his readers, science was an essentially progressive power, leading us onward and upwards as intrepid and resourceful humans explored new forces, powers and places.

F.O. Morris and Verne were among those who influenced H.G. Wells in his science fiction, as well as Huxley his beau-ideal of the scientist. His first novel in this genre was *The Time Machine*, published in the year Huxley died (1895), with its gloomy vision of evolutionary degeneration that had led to Eloi and

Morlocks; and then of a dismally cooling world running down.[48] Wells followed this in 1896 with the horrifying *The Island of Doctor Moreau*, exploring anxieties about vivisection and cross-breeding; in 1897, with *The Invisible Man*; in 1898, *The War of the Worlds*; and, in 1901, *The First Men in the Moon*. There was little comfort or assurance of progress in his novels. In 1914 he coined the slogan 'The War that will end War', but came to repent of such optimism; and at the end of his life, in 1945, he published the dismal *Mind at the End of its Tether*. His haunting pessimism about science and progress, and the example of his powerful novels, made him extremely influential.

Gloom, and dread of an impending apocalypse, did not come into mainstream science with the advent of twentieth-century global warming, but was a feature of the 'Social Darwinism' of Wells' day.[49] That is a heresy to us, though we toy happily enough with evolutionary psychology and socio-biology. Nature was seen as both a model and a threat. There were a range of 'Social Darwinisms', flourishing in different countries, not all of them politically on the right; but the principle was to not to interfere as natural selection worked on human populations. Darwin had owed a crucial insight to the social science of Malthus; now his theory was paying back its debt. Darwin himself had feared that his unfitness, ill-health, would be passed on to his children, especially because of the inbreeding of Darwins and Wedgwoods; but his disciples were chiefly concerned about the underclass, degenerates, throwbacks to our bestial origins, born criminals.[50] Particularly prominent was Cæsar Lombroso, who used statistics and data collected from convict populations on which to base his *The Female Offender* (1895) and other works.[51] Left to themselves, degenerate families would not survive and breed; but welfare states would preserve them, and their progeny would swamp the improved offspring of small middle-class families. The eugenic movement, notably promoted by Galton, flowed from this crude model. Putting socio-political problems into biological dress may have improved or confused public understanding; certainly it meant that science was perceived as very important in life, and not just as the basis of technology.

Sherlock Holmes, the model of cool rationality despite his strange habits, became the best-known fictional scientist of the last years of the nineteenth century, to the annoyance of Conan Doyle, his creator. Doyle himself became notorious for his belief in fairies and in spiritualism; and the turn of the century was a time of great enthusiasm for all sorts of occult beliefs. These went, curiously enough, with a devotion to modernism; and the ubiquity of magical beliefs is a surprise after all the popular science that we have met. But the Order of the Golden Dawn, theosophy, experiments with drugs, and sexual magic were as typical of the times as the physics of the Cavendish Laboratory.[52]

We have followed public understanding of science from the French Revolution to the First World War, and found it to be a curious business. Victorians laughed at the credulity of their times, about alternative medicine, spiritualism and superstition. Theirs was the first age of science, the second scientific revolution.[53] As Davy had forecast, prosperity and health in Europe and North America came to be based upon science; and yet few understood it in any detail, and at times distaste for it was prominent. Science became a profession, with its certified experts. It clashed with other priorities, as in the dispute about Kew Gardens between those who saw it primarily as scientific, and others for whom it was recreational (and should be cheaper and more accessible).[54] There were different publics, needing to understand at different levels and for different purposes.

Twentieth-century science was rather different. It lost its innocence in the first World War, and it became later in the century an extremely expensive business, where having the latest toys was crucial.[55] When Davy wanted a super battery to continue his researches, and heard that Napoleon was funding one for the French Institut, he appealed to the members of the Royal Institutions, skilfully playing upon institutional pride and patriotism. By the late twentieth century, drafting successful applications for funds had become an essential attribute of the successful scientist. Again, nineteenth-century science was essentially a business of individuals: there were some joint papers, but almost all had a single author. In the twentieth century, science was done by groups, and long lists of names (not all of whom may have played a very large role) adorn the heads of scientific papers.

Nineteenth-century industry increasingly required trained scientists and engineers, though there had been widespread suspicion of airy-fairy outsiders who understood little of actual processes[56] – rather as outside 'consultants' are widely regarded today. But the works laboratory was in place in many industries by the early twentieth century, and here as well as quality control there was increasing provision for research. Sometimes this might be surprisingly open-ended; and industrial laboratories, especially as equipment became more and more expensive, became great centres of science. Industrial sponsorship of university research, and then, especially with the Second World War, government sponsorship, brought the necessary funds, but also came with strings attached.[57] The idea of science as public knowledge, and its associated hopes of public understanding, was hard to maintain in a world of patents and official secrets.

Clearly, the war on ignorance which is how popularisers of science would like to see their activity, is never won. As the world changes, so does what is required to make science understood by the various publics that make up society. We should not fancy ourselves wiser than our ancestors: indeed, we need to learn from them, and their successes and failures. We have astrology, diet fads, pseudo-scientific advertising for beauty products and drugs in the snake-oil tradition, sects, and fundamentalisms. Clearly, public understanding of science has left much to be desired. We tend to take technology for granted, and find science boring – we live in a culture of suspicion, and rejection of authority, where journalistic point-scoring rates above serious investigation. But science has never had it very easy; the boffin has always been for many a figure of fun. There is little room in science nowadays for the gifted and witty amateur. Science is like other things: we shall never understand it, a practical, social and intellectual activity, unless we are aware of its history. Its boundaries are not absolute and given; its combinations of craft and reason make it fascinating, and certainly the best guide we have got to the natural world. It is even fun. Linus Pauling, one of the greatest of twentieth-century scientists, wrote:[58]

> Chemistry is wonderful. I feel sorry for people who don't know anything about chemistry. They are missing an important source of happiness.

That's the spirit! And let us hope that future publics will feel that way, and not simply get solemn messages about how important and difficult science is, rather than how exciting.

Notes

1 Understanding

1 T. Sorell, *Scientism; Philosophy and the Infatuation with Science*, London: Routledge, 1991.
2 D.M. Knight, *Humphry Davy: Science and Power*, 2nd edn, Cambridge: Cambridge University Press, 1998, p. 42.
3 D. Sobel, *Longitude*, London: Fourth Estate, 1995; C. Burr, *The Emperor of Scent: a Story of Perfume, Obsession, and the Last Mystery of the Senses*, London: Heinemann, 2003.
4 M. Faraday, 'A Speculation Touching Electric Conduction and the Nature of Matter', and 'Thoughts on Ray Vibrations', *Experimental Researches in Electricity*, London: Taylor and Francis, 1839–55, II, pp. 284–93, III, pp. 447–52.
5 T.H. Levere, *Poetry Realised in Nature: Samuel Taylor Coleridge and Early 19th-century Science*, Cambridge: Cambridge University Press, 1981.
6 S. T. Coleridge, *Hints Towards the Formation of a More Comprehensive Theory of Life*, S. B. Watson (ed.), London: Churchill, 1843; R.J. Richards, *The Romantic Conception of Life: Science and Philosophy in the Age of Goethe*, Chicago, IL: Chicago University Press, 2002.
7 H. Hellman, *Great Feuds in Science*, and *Great Feuds in Technology*, New York, NY: Wiley, 1998 and 2004; I. Hargittai, *Candid Science: Conversations with Famous Chemists*, London: Imperial College, 2000; J.A. Secord, *Controversy in Victorian Geology: the Cambrian–Silurian Dispute*, Princeton, NJ: Princeton University Press, 1986.
8 M.W. Travers, *A Life of Sir William Ramsay*, London: Arnold, 1956, pp. 172, 175–82, 188.
9 D.M. Knight, *The Age of Science: the Scientific World-View in the Nineteenth Century*, 2nd edn, Oxford: Blackwell, 1988.
10 J.T. Merz, *A History of European Thought in the Nineteenth Century* [1904–12], New York, NY: Dover, 1965.
11 M.P. Crosland, *Science Under Control: the French Academy of Sciences, 1795–1914*, Cambridge: Cambridge University Press, 1992.
12 W.A. Smeaton, *Fourcroy: Chemist and Revolutionary, 1755–1809*, Cambridge: Heffer, 1962.
13 J.P.R. Deleuze, *History and Description of the Royal Museum of Natural History*, Paris: Royer, 1823.
14 Frank James' Presidential Lecture on the lamp will be published in the *Transactions of the Newcomen Society*.
15 D.M. Knight, *Science and Spirituality: the Volatile Connection*, London: Routledge, 2004, p. 112.

16 A.G. Duff (ed.), *The Life-work of Lord Avebury (Sir John Lubbock) 1834–1913*, London: Watts, 1924, p. 80.
17 J. Issitt, *Jeremiah Joyce: Political Radical and Science Educator*, Aldershot: Ashgate, 2006.
18 J.S. Mill, *Autobiography* [1873], H. Laski (ed.), Oxford: Oxford University Press, 1963, p. 14.
19 J.B. Morrell, 'The Chemist Breeders: the Research Schools of Liebig and Thomas Thomson', *Ambix*, 19 (1972), 1–46.
20 H. Bence Jones, *The Life and Letters of Faraday*, 2nd edn, London: Longman, 1870, vol. 1, p. 11.
21 W.H. Brock and A.J. Meadows, *The Lamp of Learning: Taylor and Francis and the Development of Science Publishing*, London: Taylor and Francis, 1984.
22 On these and other people, see B. Lightman (ed.), *The Dictionary of Nineteenth-century British Scientists*, 4 vols, Bristol: Thoemmes, 2004: a very valuable guide, including publishers, editors and popularisers.
23 D. Edgerton, *Science, Technology, and the British Industrial 'Decline' 1870–1970*, Cambridge: Cambridge University Press, 1996; E. Homburg, 'Two Factions, One Profession', in D.M. Knight and H. Kragh (eds), *The Making of the Chemist: the Social History of Chemistry in Europe, 1798–1914*, Cambridge: Cambridge University Press, 1998, pp. 39–76; E. Homburg, A.S. Travers, and H.G. Schrötter (eds), *The Chemical Industry in Europe, 1850–1914: Industrial Growth, Pollution and Professionalization*, Dordrecht: Kluwer, 1998.
24 C.P. Snow, *The Two Cultures and the Scientific Revolution*, Cambridge: Cambridge University Press, 1959.
25 P. Ackroyd, *Blake*, London: Sinclair-Stevenson, 1995, p. 88.
26 K.A. Neeley, *Mary Somerville: Science, Illumination, and the Female Mind*, Cambridge: Cambridge University Press, 2001.
27 P. Metcalf, *James Knowles: Victorian Editor and Architect*, Oxford: Oxford University Press, 1980.
28 B. Lightman, 'Fighting even with death: Balfour, Scientific Naturalism, and Thomas Henry Huxley's final battle', in A.P. Barr (ed.), *Thomas Henry Huxley's Place in Science and Letters*, Athens, GA: Georgia University Press, 1997, pp. 323–50; D.M. Knight, *Science and Spirituality: the Volatile Connection*, London: Routledge, 2004, pp. 125–9.
29 J. van Wyhe, *Phrenology and the Origins of Victorian Scientific Naturalism*, Aldershot: Ashgate, 2004.
30 K. von Reichenbach, *Researches on Magnetism, Electricity, Heat, Light, Crystallization, and Chemical Attraction, in their Relations to the Vital Force*, tr. W. Gregory, London: Taylor, Walton and Gregory, 1850.
31 K. Von Reichenbach, *Lectures on Od and Magnetism* [1852], tr. F.D. O'Byrne, London: Hutchinson, 1926.
32 J. Oppenheim, *The Other World: Spiritualism and Psychical Research in England, 1850–1914*, Cambridge: Cambridge University Press, 1985.
33 See the special issue, 'Colonial Science', *Isis*, 96 (2005), 52–87.
34 W. Kaye Lamb (ed.), *The Journals and Letters of Sir Alexander Mackenzie*, Cambridge: Hakluyt Society, 1970.
35 J. Banks, *The Endeavour Journal, 1768–71*, J.C. Beaglehole (ed.), Sydney: Angus and Robertson, 1962.
36 See the series of facsimile reprints, *Scientific Travellers, 1789–1874*, 9 vols, D.M. Knight (ed.), London: Routledge and the Natural History Museum, 2003.
37 This, and Wallace's *Natural Selection* [1870], are among the books reprinted in facsimile in D.M. Knight (ed.), *The Evolution Debate, 1813–1870*, 9 vols, London: Routledge and the Natural History Museum, 2003.
38 J.F.W. Herschel, *A Manual of Scientific Enquiry: Prepared for the Use of Officers in*

Her Majesty's Navy; and Travellers in General, London: Murray, 1849; 1851 edn, D.M. Knight (ed.), Folkestone: Dawson, 1974.
39 W. Paley, *Natural Theology* [1802], M. Eddy and D.M. Knight (eds), Oxford: Oxford University Press, 2006.
40 1 Corinthians 1, 30; D.E. Ford and G. Stanton, *Reading Texts, Seeking Wisdom*, London: SCM, 2003, p. 10.

2 God's clockworld

1 J.H. Brooke, *Science and Religion: Some Historical Perspectives*, Cambridge: Cambridge University Press, 1991; and G. Cantor, *Reconstructing Nature: the Engagement Between Science and Religion*, Edinburgh: T. & T. Clark, 1998; D.C. Lindberg and R.L. Numbers (eds), *Where Science and Christianity Meet*, Chicago: Chicago University Press, 2003.
2 M. Compañon, *La obra sobre Trujillo del Peru en el siglo XVIII*, Madrid: Cultura Hispanica, 1978.
3 D.M. Knight, *Science and Spirituality: the Volatile Connection*, London: Routledge, 2004.
4 J. Priestley, *Disquisitions Relating to Matter and Spirit*, London: Johnson, 1777.
5 H.B. Carter, *Sir Joseph Banks 1743–1820*, London: Natural History Museum, 1988.
6 J. Gascoigne, *Joseph Banks and the English Enlightenment*, Cambridge: Cambridge University Press, 1994; and *Science in the Service of Empire: Joseph Banks, the British State and the Uses of Science in the Age of Revolutions*, Cambridge: Cambridge University Press, 1998.
7 T. de Quincey, *Confessions of an English Opium-eater, and Autobiography*, E.Sackville-West (ed.), London: Cresset, 1950, pp. 72–6.
8 C.U.M. Smith and R. Arnott (eds), *The Genius of Erasmus Darwin*, Aldershot: Ashgate, 2005.
9 J. Uglow, *The Lunar Men*, London: Faber and Faber, 2002.
10 D. King-Hele, *Erasmus Darwin: a Life of Unequalled Achievement*, London: de la Mare, 1999.
11 W. Wordsworth and S.T. Coleridge, *Lyrical Ballads* [1798], M. Schmidt (ed.), London: Penguin, 1999.
12 A. Desmond and J. Moore, *Darwin*, London: Penguin, 1991, p. 5.
13 O. Mayr, *Authority, Liberty and Automatic Machinery in Early Modern Europe*, Baltimore: Johns Hopkins University Press, 1986.
14 R. Porter, *Enlightenment: Britain and the Creation of the Modern World*, London: Penguin, 2000.
15 D. Sobel, *Longitude*, London: Fourth Estate, 1995.
16 W. Paley, *Natural Theology* [1802], M.D. Eddy and D.M. Knight (eds), Oxford: Oxford University Press, 2006.
17 W. Hague. *William Pitt the Younger*, London: HarperCollins, 2004, p. 292.
18 M.P. Crosland, *Science Under Control*, Cambridge: Cambridge University Press, 1992.
19 D.M. Knight, *Humphry Davy: Science and Power*, 2nd edn, Cambridge: Cambridge University Press, 1998.
20 D.M. Knight, 'Higher Pantheism', *Zygon*, 35 (2000), 603–12.
21 A. Giekie, *The Life of Sir Roderick I. Murchison*, London: Murray, 1875, vol. 1, p. 94.
22 J.J. Tobin, *Journal of a Tour made in the Years 1828–1829 Through Styria, Carniola, and Italy, Whilst Accompanying the Late Sir Humphry Davy*, London: Orr, 1832.
23 J.R. Topham, '"Beyond the Common Context": the Production and Reading of the Bridgewater Treatises', *Isis*, 89 (1998), 233–62.
24 F.M. Turner, *John Henry Newman: the Challenge to Evangelical Religion*, New

Haven, CT: Yale University Press, 2002; S. Gilley, *Newman and his Age*, London: Darton, Longman and Todd, 1990.
25 C.D. Andriesse, *Titan: a Biography of Christiaan Huygens*, tr. S. Miedema, Utrecht: Utrecht University Press, 2003.
26 M. Fisch and S. Schaffer (eds), *William Whewell: a Composite Portrait*, Oxford: Oxford University Press, 1991; R. Yeo, *Defining Science: William Whewell, Natural Knowledge and Public Debate in Early Victorian Britain*, Cambridge: Cambridge University Press, 1993; M. Fisch, *William Whewell, Philosopher of Science*, Oxford: Oxford University Press, 1991.
27 M.J.S. Rudwick, *Scenes from Deep Time: Early Pictorial Representations of the Prehistoric World*, Chicago, IL: Chicago University Press, 1992.
28 E.O. Gordon, *The Life and Correspondence of William Buckland*, London: Murray, 1894; D.M. Knight, *Science and Spirituality: the Volatile Connection*, London: Routledge, 2004, pp. 53–73.
29 C. Lyell, *Principles of Geology* [1830–3], intr. M.J.S. Rudwick, Lehre: Cramer, 1970.
30 E. Darwin, *A Century of Letters*, H. Litchfield (ed.), London: Murray, 1915, vol. 2, p. 96.
31 This is reprinted in facsimile in D.M. Knight (ed.), *The Evolution Debate, 1813–1870*, vols 2 and 3, London: Routledge and Natural History Museum, 2003.
32 L. Agassiz, *Studies on Glaciers* [1840], tr. A.V. Carozzi, New York, NY: Hafner, 1967.
33 M. Rudwick, *Scenes from Deep Time*, Chicago, IL: Chicago University Press, 1992; A. Tennyson, 'In Memoriam', S. Shatto and M. Shaw (eds), Oxford: Oxford University Press, 1982, p. 80.
34 L. Borley (ed.), *Celebrating the Life and Times of Hugh Miller: Scotland in the Early Nineteenth Century*, Cromarty: Cromarty Arts Trust, 2003.
35 R. Chambers, *Vestiges of the Natural History of Creation, and other Evolutionary Writings*, J. Secord (ed.), Chicago, IL: Chicago University Press, 1994.
36 K.E. von Baer, 'Philosophical Fragments', *Scientific Memoirs*, 6 [Natural History], (1853), 186–238.
37 J. Secord, *Victorian Sensation: the Extraordinary Publication, Reception, and Secret Authorship of Vestiges of the Natural History of Creation*, Chicago, IL: Chicago University Press, 2000.
38 A. Desmond, *The Politics of Evolution: Morphology, Medicine, and Reform in Radical London*, Chicago, IL: Chicago University Press, 1992.
39 W.J. Astore, *Observing God: Thomas Dick, Evangelicalism, and Popular Science in Victorian Britain and America*, Aldershot: Ashgate, 2001.
40 S. Parkes, *The Chemical Catechism*, 4th edn, London: Lackington Allen, 1810.
41 T. Fulford, D. Lee and P.J. Kitson, *Literature, Science and Exploration in the Romantic Era*, Cambridge: Cambridge University Press, 2004 is too severe on evangelicals; S. Coleman and L. Carlin (eds), *The Cultures of Creationism: anti-Evolutionism in English-speaking Countries*, Aldershot: Ashgate, 2004.
42 A. Fyfe, *Science and Salvation: Evangelical Popular Science Publishing in Victorian Britain*, Chicago, IL: Chicago University Press, 2004.
43 W. Newman, *Promethean Ambitions: Alchemy and the Quest to Perfect Nature*, Chicago: Chicago University Press, 2004.
44 W.R. Newman, *Promethean Ambitions: Alchemy and the Quest to Perfect Nature*, Chicago, IL: Chicago University Press, 2004.
45 G. Fownes, *Chemistry as Exemplifying the Wisdom and Beneficence of God*, London: Churchill, 1844, p. 157.
46 W.H. Brock, *Justus von Liebig: the Chemical Gatekeeper*, Cambridge: Cambridge University Press, 1997.
47 J. Liebig, *Familiar Letters on Chemistry*, 3rd edn, London: Taylor, Walton, and Maberly, 1851, pp. 18–20.

48 F.W.J. Schelling, *Ideas for a Philosophy of Nature*, tr. E.E. Harris and P. Heath, intr. R. Stern, Cambridge: Cambridge University Press, 1988.
49 G.F.W. Hegel, *Philosophy of Nature*, ed. and tr. M.J. Petry, London: Allen and Unwin, 1970.
50 H.C. Oersted, *Selected Scientific Works*, tr. and ed. K. Jelved, A.D. Jackson and O. Knudsen, intr. A.D. Wilson, Princeton, NJ: Princeton University Press, 1998.
51 H.C. Oersted, *The Soul in Nature*, tr. L. and J.B. Horner, London: Bohn, 1852.
52 F. Burkhardt *et al.* (eds), *The Correspondence of Charles Darwin*, vol. 4, Cambridge: Cambridge University Press, 1988, p. 488.
53 A stimulating conference on Oersted was held at Harvard in May 2002 under the auspices of Gerald Holton, and will be published in 2006.
54 G. Wilson, *Religio Chemici: Essays*, London: Macmillan, 1862; J.A. Wilson, *Memoir of George Wilson*, Edinburgh: Edmonston and Douglas, 1860.
55 R. Anderson, '"What is Technology?": Education through Museums in the Mid-nineteenth Century', *BJHS*, 25 (1992), 169–84.
56 T. Browne, *The Works*, G. Keynes (ed.), 2nd edn, London: Faber, 1964, vol. 1.
57 W.S. Symonds, *Old Bones: or, Notes for Young Naturalists*, London: Hardwicke, 1861; D.M. Knight, 'Old Bones and New Ideas', in N. Cooper, *John Ray and his Successors: the Clergyman as Biologist*, Braintree: John Ray Trust, 2000, pp. 145–51.
58 N.A. Rupke, *Richard Owen: Victorian Naturalist*, New Haven, CT: Yale University Press, 1994.
59 R.J. Richards, *The Romantic Conception of Life: Science and Philosophy in the Age of Goethe*, Chicago, IL: Chicago University Press, 2002; E.P. Hamm, 'Romantic Life and Science', *Annals of Science*, 62 (2005), 377–85; L. Nyhart, *Biology Takes Form: Animal Morphology and the German Universities, 1800–1900*, Chicago, IL: Chicago University Press, 1995.
60 L. Oken, *Elements of Physiophilosophy*, tr. A. Tulk, London: Ray Society, 1857.
61 F. Gregory, *Nature Lost? Natural Science and the German Theological Traditions of the Nineteenth Century*, Cambridge, MA: Harvard University Press, 1992.
62 A. Tennyson, *In Memoriam* [1850], S. Shatto and M. Shaw (eds), Oxford: Oxford University Press, 1982, p. 79.
63 A. Scott, 'Visible Incarnations of the Unseen': Henry Drummond and the Practice of Typological Exegesis', *BJHS*, 37 (2004), 435–54.
64 P.H. Gosse, *Omphalos: an Attempt to Untie the Geological Knot* [1857], intr. D.M. Knight, London: Natural History Museum and Routledge, 2004; A. Thwaite, *Glimpses of the Wonderful: the Life of Philip Henry Gosse*, London: Faber, 2002.
65 G. Cantor, 'Michael Faraday Meets the "High-Priestess of God's Works": a Romance on the Theme of Science and Religion', in D.M. Knight and M.D. Eddy (eds), *Science and Beliefs*, Aldershot: Ashgate, 2005, pp. 157–70.

3 Holding forth

1 B.C. Brodie, *Autobiography*, London: Longman, 1865, p. 61, but see also pp. 78, 80, 128–30, 155.
2 *Register of Arts and Sciences*, 1 (1824), 3–5; and see D.M. Knight, 'Scientific Lectures: a History of Performance', *Interdisciplinary Science Reviews*, 27 (2002), 217–24, and P. Bertucci and G. Pancaldi (eds), *Electric Bodies: Episodes in the History of Medical Electricity*, Bologna: Dipartimento di Filosofia, Universitá di Bologna, 2001.
3 J. Browne, *Charles Darwin: Voyaging*, New York: Knopf, 1995.
4 H. Lonsdale, *A Sketch of the Life and Writings of Robert Knox, the Anatomist*, London: Macmillan, 1870.
5 C.A. Russell, 'Richard Watson: Gaiters and Gunpowder', in M. Archer and C. Haley

(eds), *The 1702 Chair of Chemistry at Cambridge: Transformation and Change*, Cambridge: Cambridge University Press, 2005, pp. 57–83.
6. R. Watson, *Chemical Essays*, 6th edn, London: Evans, 1793.
7. P.C. Almond, *Adam and Eve in 17th-century Thought*, Cambridge: Cambridge University Press, 1999.
8. L. Colley, *Britons: Forging the Nation, 1707–1837*, New Haven, CN: Yale University Press, 1992.
9. A.J. La Vopa, *Grace, Talent, and Merit: Poor Students, Clerical Carers, and Professional Ideology in 18th-century Germany*, Cambridge: Cambridge University Press, 1988.
10. J. Uglow, *The Lunar Men*, London: Faber and Faber, 2002.
11. F.A.J.L. James (ed.), *The Common Purposes of Life: Science and Society at the Royal Institution*, Aldershot: Ashgate, 2002: M. Berman, *Social Change and Scientific Organization: the Royal Institution, 1799–1844*, London: Heinemann, 1978.
12. H.B. Carter, *Sir Joseph Banks,1743–1820*, London: Natural History Museum, 1988; J.C. Beaglehole (ed.), *The Endeavour Journal of Sir Joseph Banks*, Sydney: Angus and Robertson, 1962.
13. J. Gascoigne, *Science in the Service of Empire: Joseph Banks, the British State and the Uses of Science in the Age of Revolution*, Cambridge: Cambridge University Press. 1998, p. 4.
14. N.H. Robinson and E.G. Forbes, *The Royal Society Catalogue of Portraits*, London: Royal Society, 1980, p. 18.
15. S.C. Brown (ed.), *The Collected Works of Count Rumford*, 5 vols, Cambridge, MA: Harvard University Press, 1968–70.
16. T. Garnett, *Outlines of a Course of Lectures on Chemistry*, London: Cadell and Davies, 1801.
17. T. Young, *A Course of Lectures on Natural Philosophy and the Mechanical Arts*, London: Johnson, 1807.
18. C. Jungnickel and R. McCormmach, *Cavendish*, Philadelphia PA: American Philosophical Society, 1996; F. Seitz, 'Henry Cavendish: the Catalyst for the Chemical Revolution', *Notes and Records of the Royal Society*, 59 (2005), 175–99; see also pp. 215–18.
19. A.L. Barbauld, *The Poems*, W. McCarthy and E. Kraft (eds), Athens, GA: Georgia University Press, 1994, p. 158.
20. J. Chrichton-Browne, 'John Tyndall', *Proceedings of the Royal Institution*, 14 (1893–5), pp. 161–8, esp. p. 162; Lord Rayleigh, 'The Scientific Work of Tyndall', same issue, pp. 216–24.
21. See the special issue of *Interdisciplinary Science Reviews*, 27 (2002), 161–247.
22. M. Bresadola and G. Pancaldi (eds), *Luigi Galvani International Workshop: Proceedings*, Bologna: Dipartimento di Filosofia, Universitá di Bologna, 1999; M. Beretta and K. Grandin (eds), *A Galvanized Network: Italian–Swedish Scientific Relations from Galvani to Nobel*, Stockholm: Royal Swedish Academy of Sciences, 2001.
23. J. Marcet, *Conversations on Chemistry*, 11th edn, London: Longman, 1828, vol. 1, p. v; Davy was Professor when the first edition was published, and the wording in the original remained unchanged.
24. H. Davy, *Collected Works*, J. Davy (ed.), London: Smith Elder, 1839–40, vol. 8, p. 313.
25. W. Shakespeare, *The Tempest*, IV, 1, 151.
26. J.J. McGann, *The New Oxford Book of Romantic Period Verse*, Oxford: Oxford University Press, 1993, pp. 531–2.
27. M.P. Crosland, *Science Under Control: the French Academy of Sciences, 1795–1914*, Cambridge: Cambridge University Press, 1992.
28. J.Z. Fullmer, *Humphry Davy's Published Works*, Cambridge, MA: Harvard University Press, 1969, p. 95.

29 C. Babbage, *Reflections on the Decline of Science in England, and on Some of its Causes*, London: Fellowes, 1830, pp. 105–8; M.V. Wilkes, 'Charles Babbage and his World', *Notes and Records of the Royal Society*, 56 (2002), 353–65.
30 T. Griffiths, *Chemistry of the Four Ancient Elements, Fire, Air, Earth and Water*, London: Highley, 1842.
31 J. Scoffern, *Chemistry no Mystery: or, a Lecturer's Bequest*, 2nd edn, London: Hall [1848].
32 J. Morrell, *John Phillips and the Business of Victorian Science*, Aldershot: Ashgate, 2005, pp. 127–8.
33 J.G. Spurzheim, *The Physiognomical System of Drs Gall and Spurzheim; Founded on Anatomical and Physiological Examination of the Nervous System in General, and of the Brain in Particular; and Indicating the Dispositions and Manifestations of the Mind*, 2nd edn, London: Baldwin, Cradock and Joy, 1815; J. van Wyhe, *Phrenology and the Origins of Victorian Scientific Naturalism*, Aldershot: Ashgate, 2004.
34 W. St. Clair, *The Reading Nation in the Romantic Period*, Cambridge: Cambridge University Press, 2004.
35 *Fourth Report of the Royal Commission on Scientific Instruction and the Advancement of Science*, London: HMSO, 1874, pp. 18–24; R. Anderson, '"What is Technology?": Education through Museums in the Mid-19th Century', *BJHS*, 25 (1992), 169–84.
36 P. White, *Thomas Huxley: Making the 'Man of Science'*, Cambridge: Cambridge University Press, 2003, pp. 148–51; A. Desmond, *Huxley: Evolution's High Priest*, London: Michael Joseph, 1997, pp. 256–61.
37 A.P. Barr (ed.), *The Major Prose of Thomas Henry Huxley*, Athens, GA: Georgia University Press, 1997, pp. 154–73; quotations from p. 172. See also the accompanying essays, A.P. Barr (ed.), *Thomas Henry Huxley's Place in Science and Letters*, Athens, GA: Georgia University Press, 1997.
38 B. Lightman, *The Origins of Agnosticism*, Baltimore, MD: Johns Hopkins University Press, 1987; A. Pyle (ed.), *Agnosticism: Contemporary Responses to Spencer and Huxley*, Bristol: Thoemmes, 1995.
39 *Royal Commission on Scientific Instruction and the Advancement of Science*, London: HMSO, 1872, vol. 1, appendix, pp. 67–70 for an index of his evidence, and paragraph 3,640 for opposition to overspecialism.
40 H.L. Mansel, *The Limits of Religious Thought Examined*, 4th edn, London: Murray, 1859.
41 B. Jowett *et al.*, *Essays and Reviews*, London: Parker, 1860.

4 Poetry, metaphor and algebra

1 W.T. Stearn, *Botanical Latin*, 2nd edn, Newton Abbot: David and Charles, 1973.
2 J. Ruskin, *Proserpina: Studies of Wayside Flowers*, Orpington: Allen, 1879, p. 6.
3 D. King-Hele, *Erasmus Darwin: a Life of Unequalled Achievement*, London: de la Mare, 1999.
4 J. Uglow, *The Lunar Men: the Friends who made the Future*, London: Faber and Faber, 2002.
5 E. Darwin, *The Botanic Garden, Part Two Containing the Loves of the Plants* [1789–91], Menton: Scolar Press, 1973, vol. 2, 'advertisement' and pp. 102–3.
6 E. Darwin, *The Temple of Nature; or, the Origin of Society* [1803], Mnton: Scolar, 1973, pp. 1, 139.
7 G.S. Kirk and J.E. Raven, *The Pre-Socratic Philosophers: a Critical History with a Selection of Texts*, Cambridge: Cambridge University Press, 1962, pp. 320–61.
8 D.M. Knight, *Science and Spirituality: the Volatile Connection*, London: Routledge, 2004, pp. 94–7, 106.

204 Notes

9 W. Gifford (ed.), *Poetry of the Anti-Jacobin*, London: Wright, 1799, pp. 108–29, 134–41; W. Hague, *William Pitt the Younger*, London: HarperCollins, 2004, p. 418.
10 W.R. Newman, *Promethean Ambitions: Alchemy and the Quest to Perfect Nature*, Chicago, IL: Chicago University Press, 2004.
11 L.M. Principe and L. De Witt, *Transmutations: Alchemy in Art*, Philadelphia, PA: Chemical Heritage, 2002.
12 D.M. Knight, *Ideas in Chemistry: a History of the Science*, London: Athlone, 1992, pp. 13–26.
13 C.D. Patrides (ed.), *The English Poems of George Herbert*, London: Dent, 1974.
14 C.G. Jung, *Psychology and Alchemy*, tr. R.F.C. Hull, 2nd edn, London: Routledge, 1968; see L.M. Principe and W.R. Newman, 'Some Problems with the Historiography of Alchemy', in W.R. Newman and A. Grafton (eds), *Secrets of Nature: Astrology and Alchemy in Early Modern Europe*, Cambridge, MA: MIT, 2002, pp. 385–431; A.G. Debus (ed.), *Alchemy and Early Modern Chemistry: Papers from Ambix*, London: SHAC, 2004.
15 D. Freedberg, *The Eye of the Lynx: Galileo, his Friends, and the Beginnings of Modern Natural History*, Chicago, IL: Chicago University Press, 2002, p. 194; and cf. pp. 336, 344 for empirical evidence leading to a false conclusion.
16 P.J. Macquer, *Elements of the Theory and Practice of Chymistry*, tr. Andrew Reid, 3rd edn, London: Nourse, 1775, vol. 1, pp. viii, ix.
17 P.J. Macquer, *Dictionnaire de Chymie*, 2nd edn, 4 vols, Paris: Barrois, 1778.
18 G. de Morveau, A.L. Lavoisier, C.L. Berthollet, and A.F. Fourcroy, *Méthode de Nomenclature Chimique*, Paris: Cuchet, 1787.
19 J. Simon, *Chemistry, Pharmacy and Revolution*, Aldershot: Ashgate, 2004.
20 J. Golinski, *Science as Public Culture: Chemistry and Enlightenment in Britain, 1760–1820*, Cambridge: Cambridge University Press, 1992, pp. 11–49, 190–202.
21 A.L. Lavoisier, *Elements of Chemistry*, tr. R. Kerr, Edinburgh: Creech, 1790, pp. xxx, xxxii.
22 M. Chaouli, *The Laboratory of Poetry: Chemistry and Poetics in the Work of Friedrich Schlegel*, Baltimore, CT: Johns Hopkins University Press, 2002.
23 C.A. Russell, 'Richard Watson: Gaiters and Gunpowder', C. Haley and P. Wothers, 'Lavoisier's Chemistry comes to Cambridge', and M. Usselman, 'Smithson Tennant: the Innovative and Eccentric Eighth Professor of Chemistry', in M. Archer and C. Haley (eds), *The 1702 Chair of Chemistry at Cambridge: Transformation and Change*, Cambridge: Cambridge University Press, 2005.
24 J. Priestley, *Experiments and Observations on Air, and other Branches of Natural Philosophy*, Birmingham: Pearson, 1790, vol. 1, pp. xvi–xviii, vol. 3, p. 563.
25 W. Nicholson, *A Dictionary of Practical and Theoretical Chemistry*, 2nd edn, London: Phillips, 1808.
26 W. Nicholson, *The First Principles of Chemistry*, 3rd edn, London: Robinson, 1796, pp. vii–ix.
27 T.S. Kuhn, 'The Function of Dogma in Scientific Research', in A.C. Crombie (ed.), *Scientific Change*, Oxford: Oxford University Press, 1963, pp. 347–69. His book, *The Structure of Scientific Revolutions*, Chicago, IL: Chicago University Press, 1962, was already in the press.
28 J. Priestley, *Heads of Lectures on a Course of Experimental Philosophy, Particularly Including Chemistry, delivered at the New College, Hackney*, London: Johnson, 1794, p. 3.
29 R.E. Schofield (ed.), *A Scientific Autobiography of Joseph Priestley*, Cambridge, MA: MIT, 1966. p. 313.
30 J. Cottle, *Reminiscences of Samuel Taylor Coleridge and Robert Southey*, London: Houlston & Stonemen, 1847.
31 R. Lamont-Brown, *Humphry Davy: Life Beyond the Lamp*, London: Sutton, 2004, pp. 37–40.

32 A. Tennyson, *In Memoriam*, S. Shatto and M. Shaw (eds), Oxford: Oxford University Press, 1982, p. 80.
33 J.A. Secord, *Victorian Sensation: the Extraordinary Publication, Reception, and Secret Authorship of Vestiges of the Natural History of Creation*, Chicago, IL: Chicago University Press, 2000.
34 L. Huxley, *Life and Letters of Thomas Henry Huxley*, 2nd edn, London: Macmillan, 1913, vol. 3, pp. 268–70.
35 *The Anti-Jacobin, or Weekly Examiner* [16 April 1798], 4th edn, London: Wright, 1799, vol. 2, pp. 170–3; E.L. de Montluzin, *The Anti-Jacobins, 1798–1800: the Eartly Contributors to the Anti-Jacobin Review*, London: Macmillan, 1988.
36 Anon., *Bubble and Squeak: a Galli-maufry of British Beef with the Chipp'd Cabbage of Gallic Philosophy and Radical Reform*, London: Wright, 1799, p. 27.
37 A. Treneer, *The Mercurial Chemist: a Life of Sir Humphry Davy*, London: Methuen, 1963, pp. 63–4.
38 H. Davy, *Collected Works*, London: Smith Elder, 1839–40, vol. 1, p. 185.
39 H. Davy, *Consolations in Travel, or the Last Days of a Philosopher*, London: Murray, 1830; D.M. Knight, *Humphry Davy: Science and Power*, 2nd edn, Cambridge: Cambridge University Press, 1996, pp. 168–83.
40 J. Davy, *Fragmentary Remains, Literary and Scientific, of Sir Humphry Davy*, London: Churchill, 1858, p. 14.
41 J. Davy, *Memoirs of the Life of Sir Humphry Davy*, London: Longman, 1836, vol. 2, p. 157.
42 D.M. Knight, 'From Science to Wisdom: Humphry Davy's Life', in M. Shortland and R. Yeo (eds), *Telling Lives*, Cambridge: Cambridge University Press, 1996, pp. 103–14.
43 J.Z. Fullmer and M. Usselman, 'Faraday's Election to the Royal Society: a Reputation in Jeopardy', *Bulletin of the History of Chemistry*, 11 (1991), 17–28.
44 J. Davy, *Memoirs of the Life of Sir Humphry Davy*, London: Longman, 1836, vol. 2, pp. 95–6, 218.
45 J.Z. Fullmer, *Young Humphry Davy: the Making of an Experimental Chemist*, Philadelphia, PA: American Philosophical Society, 2000.
46 F.A.J.L. James (ed.), *The Correspondence of Michael Faraday*, vol. 2, London: Institute of Electrical Engineers, 1993, pp. 176ff, 222, 278, 398, 463ff.
47 J.F.W. Herschel, *Preliminary Discourse on the Study of Natural Philosophy* [1830], intr. M. Partridge, New York, NY: Johnson, 1966.
48 J.F.W. Herschel, *Popular Lectures on Scientific Subjects*, London: Strahan, 1873.
49 J.F.W. Herschel, *A Treatise on Astronomy*, new edn, London: Longman, 1851, p. 5; italics Herschel's.
50 D.G. King-Hele (ed.), *John Herschel 1792–1871: a Bicentennial Commemoration*, London: Royal Society, 1992; esp. pp. 29–36, K. Moore, 'Space and Time Forgot: John Herschel as Poet'.
51 S. Ruskin, *John Herschel's Cape Voyage: Private Science, Public Imagination and the Ambitions of Empire*, Aldershot: Ashgate, 2004.
52 J.F.W. Herschel, *Essays from the Edinburgh and Quarterly Reviews, with Addresses and Other Pieces*, London: Longman, 1857, p. 737; also in D.M. Knight, *Science and Spirituality: the Volatile Connection*, London: Routledge, 2004, pp. 123–4.
53 H.C. Oersted, *Selected Scientific Works*, intr. A.D. Wilson, tr. K. Jelved, A.D. Jackson and O. Knudsen, Princeton, NJ: Princeton University Press, 1998.
54 H.C. Oersted, *The Soul in Nature*, tr. L. and J.B. Horner, London: Bohn, 1852.
55 J.F.W. Herschel, *Essays from the Edinburgh and Quarterly Reviews*, London: Longman, 1857, p. 738.
56 J.F.W. Herschel, *Essays from the Edinburgh and Quarterly Reviews*, London: Longman, 1857, p. 732; D.M. Knight, 'Words That Make Worlds', *Science and Public Affairs*, Summer 1994, pp. 23–5.

57 B. Mahon, *The Man Who Changed Everything: the Life of James Clerk Maxwell*, Chichester: Wiley, 2003.
58 L. Campbell and W. Garnet, *The Life of James Clerk Maxwell*, new edn, London: Macmillan, 1884, pp. 383–421, esp. 408, 409, 412–13.
59 J. Smith, *Fact and Feeling: Baconian Science and the 19th-century Literary Imagination*, Madison, WI: Wisconsin University Press, 1994.
60 W. Shakespeare, *A Midsummer-Night's Dream*, V, 1, 224–31.
61 B.A.A.S, *Exeter Change for the British Lions*, London: Pardon, 1869.
62 W.H. Brock (ed.), *The Atomic Debates: Brodie and the Rejection of the Atomic Theory*, Leicester: Leicester University Press, 1967.
63 A.J. Rocke, *Nationalizing Science: Adolphe Wurtz and the Battle for French Chemistry*, Cambridge, MA: MIT, 2001, p. 333.
64 R. Noakes, '*Punch* and Comic Journalism in mid-Victorian Britain', in G. Cantor, G. Dawson, G. Gooday, R. Noakes, S. Shuttleworth and J.R. Topham, *Science in the 19th-century Periodical*, Cambridge: Cambridge University Press, 2004, pp. 91–122.
65 M.J.S. Rudwick, *Scenes from Deep Time*, Chicago, IL: Chicago University Press, 1992, pp. 149, 217; A. Desmond and J. Moore, *Darwin*, London: Penguin, 1992, pl. 37.
66 *Punch*, 15 November, 1862, p. 198; 19 December 1868, p. 264.

5 Picturing science

1 J. Joyce, *Scientific Dialogues*, London: Allman, n.d.; J.S. Mill, *Autobiography*, intr. H.J. Laski, Oxford: Oxford University Press, 1924, p. 14; R. Heber, *Narrative of a Journey through the Upper Provinces of India from Calcutta to Bombay, 1824–5*, London: Murray, 1828, vol. 1, p. 365.
2 J. Issitt, 'Jane Marcet', in B. Lightman (ed.), *Dictionary of 19th-Century British Scientists*, Bristol: Thoemmes, 2004.
3 See the special issue on 'Aesthetics and Visualization in Chemistry', *Hyle*, 9 (2003), 1–243.
4 M. Faraday, *Chemical Manipulation*, 3rd edn, London: Murray, London, 1842.
5 D.M. Knight. 'Drawing the Unspeakable: Chemistry as Visual Art', *Chemistry and Industry*, 10 (1997), 384–7; '"Exalting Understanding Without Depressing Imagination": Depicting Chemical Process', *Hyle*, 9 (2003) 171–89.
6 H. Robin, *The Scientific Image: from Cave to Computer*, New York: Abrams, 1992; and the special issue, 'Illustrating Science', *Interdisciplinary Science Reviews*, 29 (2004), 225–336.
7 L.M. Principe and L. DeWitt, *Transmutations: Alchemy in Art*, Philadelphia, PA: Chemical Heritage Foundation, 2002.
8 L.M. Principe and W.R. Newman, 'Some Problems with the Historiography of Alchemy', in W.R. Newman and A. Grafton (eds), *Secrets of Nature: Astrology and Alchemy in Early Modern Europe*, Cambridge, MA: MIT, 2001, pp. 385–431.
9 D.M. Knight, *Ideas in Chemistry*, 2nd edn, London: Athlone, 1995, chapter 2; W.R. Newman, *Promethean Ambitions: Alchemy and the Quest to Perfect Nature*, Chicago, IL: Chicago University Press, 2004.
10 J. Priestley, *Experiments and Observations on Different Kinds of Air, and Other Branches of Natural Philosophy*, 3 vols, Birmingham, 1790: the plates are all in volume 1.
11 A.L. Lavoisier, *Elements of Chemistry, in a New Systematic Order, Containing all the Modern Discoveries*, trans. R. Kerr, Edinburgh: Creech, 1790.
12 M. Beretta, *Imaging a Career in Science: the Iconography of Antoine Laurent Lavoisier*, Canton, MA: Science History, 2002.
13 J.W. Goethe, *Theory of Colours*, trans. C. Eastlake, London: Murray, 1840.
14 R.P. Greg and W.G. Lettsom, *Manual of the Mineralogy of Great Britain and Ireland*, London: van Voorst, 1858.

15 A.W. Hofmann, 'On the Combining Power of Atoms', *Proceedings of the Royal Institution*, 4 (1862–6), 401–30.
16 A.M. Young, *Antique Medicine Chests: or Glyster, Blister and Purge*, London: Vernier, 1994.
17 H. Davy, *Consolations in Travel; or, The Last Days of a Philosopher*, London: Murray, 1830, pp. 250–1.
18 D.M. Knight, 'Making Chemistry Popular', *American Chemical Society Bulletin for the History of Chemistry*, 29 (2004), 1–8; relevant extracts from Marcet and Davy are reprinted in J. Hawley (general ed.), *Literature and Science*, vol. 8 *Chemistry*, B. Dolan (ed.), London: Pickering and Chatto, 2004.
19 W.T. Brande, *A Manual of Chemistry; Containing the Principal Facts of the Science, Arranged in the Order in Which They Are Discussed and Illustrated in the Lectures at the Royal Institution*, 3rd edn, London: Murray, 1830, vol. 2 frontispiece. On textbooks, see A. Lundgren and B. Bensaude-Vincent (eds), *Communicating Chemistry: Textbooks and their Audiences:* Canton, MA: Science History, 2000.
20 M. Faraday, *Curiosity Perfectly Satisfied: Faraday's Travels in Europe, 1813–15*, B. Bowers and L. Symons (eds), London: Peregrinus, 1991, pp. 23–30; H. Davy, 'Some Experiments and Observations on a New Substance Which Becomes a Violet Coloured Gas by Heat', *Philosophical Transactions*, 104 (1814), 74–93.
21 W.H. Brock, *Justus von Liebig: the Chemical Gatekeeper*, Cambridge: Cambridge University Press, 1997; E. Homburg, 'Two Factions, One Profession: the Chemical Profession in German Society, 1780–1870', in D.M. Knight and H. Kragh (eds), *The Making of the Chemist: the Social History of Chemistry in Europe, 1789–1914*, Cambridge: Cambridge University Press, 1998.
22 N. Harte and J. North, *The World of University College, London, 1828–1978*, London: UCL, 1978, p. 58.
23 *Sixth Report of the Royal Commission on Scientific Instruction and the Advancement of Science*, London: HMSO, 1875, appendix, pp. 36–43.
24 A. Lundgren and B. Bensaude-Vincent (eds), *Communicating Chemistry: Textbooks and their Audiences*, Canton, MA: Science History, 2000.
25 N. Bion, *The Construction and Principal Uses of Mathematical Instruments* [2nd edn, 1758], trans. E. Stone, facsimile, London: Holland Press, 1972.
26 H. Michel, *Scientific Instruments in Art and History*, trans. R.E.W. and F.R. Maddison, London: Barrie and Rockliff, 1967; M. Holbrook, *Science Preserved: a Directory of Scientific Instruments in Collections in the United Kingdom and Eire*, London: HMSO, 1992.
27 G. L'E. Turner, *Nineteenth-century Scientific Instruments*, London: Sotheby, 1983.
28 A.Q. Morton, *Science in the 18th Century: the King George III Collection*, London: Science Museum, 1993; with J. Wess, *Public and Private Science: the King George III Collection*, Oxford: Oxford University Press, 1993.
29 J.P. Nichol, *The Architecture of the Heavens*, London: Parker, 1850, p. ix.
30 J.A. Secord, *Victorian Sensation: the Extraordinary Publication, Reception, and Secret Authorship of Vestiges of the Natural History of Creation*, Chicago, IL: Chicago University Press, 2000, pp. 467–8; W.J. Astore, *Observing God: Thomas Dick, Evangelicalism, and Popular Science in Victorian Britain and America*, Aldershot: Ashgate, 2001.
31 M. Campbell, *David Scott, 1806–1849*, Edinburgh: National Galleries of Scotland, 1990, p. 16.
32 E. Robinson and A.E. Musson, *James Watt and the Steam Revolution: a Documentary History*, London: Adams and Dart, 1969, reproduces the steam-engine drawings in colour.
33 K. Baynes and F. Pugh, *The Art of the Engineer*, London: Lutterworth, 1981.
34 R. Hoffmann and V. Torrance, *Chemistry Imagined*, Washington, DC: Smithsonian, 1993.

35 C.C. Gillispie (ed.), *A Diderot Pictorial Encyclopedia of Trades and Industry*, New York, NY: Dover, 1959.
36 *The Illustrated Exhibitor: a Tribute to the World's Industrial Jubilee*, London: Cassell, 1851.
37 S. Forgan and G. Gooday, 'Constructing South Kensington: the Buildings and Politics of T.H. Huxley's Working Environments', *BJHS*, 29 (1996), 435–68.
38 J. Simmons, *The Victorian Railway*, London: Thames and Hudson, 1991.
39 J.C. Bourne, *Drawings of the London and Birmingham Railway*, London: Bogue, 1839; *The History and Description of the Great Western Railway*, London: Bogue, 1846.
40 L. James, *A Chronology of the Construction of Britain's Railways, 1778–1855*, London: Ian Allan, 1983.
41 A.B. Clayton, *Views on the Liverpool and Manchester Railway*, Liverpool: Cannell, 1831; T.T. Bury, *Coloured Views on the Liverpool and Manchester Railway*, London; Ackermann, 1831; J.W. Carmichael, *Views on the Newcastle and Carlisle Railway*, Newcastle: Currie and Bowman, 1836–8; A.F. Tait, *Views on the Leeds and Manchester Railway*, London: Bradshaw, 1845.
42 F.D. Klingender, *Art and the Industrial Revolution*, A. Elton (ed.), London: Evelyn, Adams & Mackay, 1968.
43 T.H. Hair, *A Series of Views of the Collieries in the Counties of Northumberland and Durham*, London: Madden, 1844.
44 C. Babbage, *Reflections on the Decline of Science in England*, London: Fellowes, 1830, p. 104.
45 M. Rudwick, *Scenes from Deep Time*, Chicago, IL: Chicago University Press, 1992.
46 C.E. Jackson, *Bird Illustrators: Some Artists in Early Lithography*, London: Witherby, 1975; *Wood Engravings of Birds*, London: Witherby, 1978; *Bird Etchings: the Illustrators and their Books, 1655–1855*, Ithaca, NY: Cornell University Press, 1985.
47 W. St. Clair, *The Reading Nation in the Romantic Period*, Cambridge: Cambridge University Press, 2004, pp. 182–5.
48 W. Savage, *Dictionary of the Art of Printing* [1841], New York, NY: Franklin, n.d., pp. 249–61.
49 W. Blunt, *The Art of Botanical Illustration*, London: Collins, 1950; D.M. Knight, *Zoological Illustration: an Essay towards a History of Printed Zoological Pictures*, Folkestone: Dawson, 1977.
50 D. Hart-Davis, *Audubon's Elephant*, London: Weidenfeld & Nicolson, 2003.
51 G.C. Sauer, *John Gould, the Bird Man: a Chronology and Bibliography*, London: Sotheran, 1982.
52 A.M. Lysaght, *The Book of Birds*, London: Phaidon, 1975; A.M. Coates, *The Book of Flowers*, London: Phaidon, 1973.
53 D.M. Knight, *Science in the Romantic Era*, Aldershot: Ashgate Variorum, 1998, pp. 197–224.
54 W. Swainson, *Ornithology: Flycatchers*, Edinburgh: Lizars, 1838.
55 C. Knight, *The Old Printer and the Modern Press*, London: Murray, 1854, pp. 260–7.
56 A. Newton, *A Dictionary of Birds*, London: Black, 1893–6, pp. ii–iii.
57 W. Swainson, *The Natural History and Classification of Birds*, London: Longman, 1836, vol. 1, frontispiece and pp. 282–5; the circles are on pp. 303, 318.
58 D.M. Knight, *Science and Spirituality: the Volatile Connection*, London: Routledge, 2004, pp. 74–82.
59 E. Lear, *Illustrations of the Family of Psittacidæ or Parrots*, London: Lear, 1832.
60 T.C. Jerdon, *Illustrations of Indian Ornithology*, Madras: Baptist Press, 1847.
61 H. Santapau (ed.), *Icones Roxburghianae, or Drawings of Indian Plants*, Calcutta: Botanical Survey of India, 1964–.
62 P.J.P. Whitehead and P.I. Edwards (eds), *Chinese Natural History Drawings Selected from the Reeves Collection*, London: British Museum (Natural History), 1974.

63 E. Perez-Arbelaez *et al.* (eds), *La Real Expedicion Botanica del Nuevo Reino de Granada*, Madrid: Instituto de Cultura Hispanica, 1954–.
64 F.E. Beddard, *Animal Coloration*, London: Swan Sonnenschein, 1892, is a classic work.
65 R. McGowan, 'The Bank of England and the Policing of Forgery', *Past and Present*, 186 (2005), 81–116.
66 G.C. Sauer, *John Gould, the Bird Man*, London: Sotheran, 1982.
67 J.S. Miller, *A Natural History of the Crinoidea, or Lily-stalked Animals*, Bristol: Ford, 1821; D.M. Knight, *Zoological Illustration: an Essay Towards a History of Printed Zoological Pictures*, Folkestone: Dawson, 1977, p. 30.
68 J.C. Prichard, *The Natural History of Man*, 3rd edn, London, Bailliere, 1848.
69 E.O. Gordon, *The Life and Correspondence of William Buckland*, London: Murray, 1894, pp. 61, 145, and more caricatures in the book; D.M. Knight, *Zoological Illustration*, Folkestone: Dawson, 1977, p. 183.
70 W.H. Helfand, *Quack, Quack, Quack: the Sellers of Nostrums in Prints, Posters, Ephemera and Books*, New York, NY: Grolier Club, 2002.
71 J.A. Paris, *The Life of Sir Humphry Davy*, London: Colburn and Bentley, 1831; H. Falconer, *Palæontological Memoirs and Notes*, C. Murchison (ed.), London: Hardwicke, 1868.
72 J. Browne, *Charles Darwin: Volume 2, the Power of Place*, London: Pimlico, 2003.
73 F. Burkhardt *et al.* (eds), *The Correspondence of Charles Darwin*, Cambridge: Cambridge University Press, 1985–.
74 J.C. Lavater, *Essays on Physiognomy: Designed to Promote the Knowledge and the Love of Mankind*, tr. T. Holcroft, 10th edn, London: Tegg, 1858; S.R. Wells, *How to Read Character: a New Illustrated Handbook of Phrenology and Physiology* [1869], Rutland, Vermont: Tuttle, 1971; J. van Wyhe, *Phrenology and the Origins of Victorian Scientific Naturalism*, Aldershot: Ashgate, 2004.
75 E.R. Tufte, *The Visual Display of Quantitative Information*, Cheshire, CN: Graphics Press, 1983.
76 L. Agassiz, *Etudes sur les Glaciers*, Neuchatel: Nicolet, 1840.
77 Cf. H.C. Escher von der Linth, *Views and Panoramas of Switzerland, 1780–1820*, Zurich: Atlantis, 1975.
78 L.A.J. Quetelet, *A Treatise on Man and the Development of his Faculties*, Edinburgh: Chambers, 1842.
79 J.F.W. Herschel, *A Treatise on Astronomy*, London: Longman, new edn, 1851, p. 5.

6 Ballyhoo

1 F. Bacon, *The Advancement of Learning, and New Atlantis*, intr. T. Case, Oxford: Oxford University Press, 1960, p. 288.
2 J. Moxon, *Mechanick Exercises: or the Doctrine of Handy-Works Applied to the Arts of Smithing, Joinery, Carpentry, Turning, Bricklaying* [3rd edn, 1703], New York, NY: Praeger, 1970; *Mechanick Exercises on the Whole Art of Printing* [1683], 2nd edn, H. Davis and H. Carter (eds), London: Oxford University Press, 1962.
3 T. Sprat, *History of the Royal Society* [1667], J.I. Cope and H.W. Jones (eds), London: Routledge, 1959, pp. 311–8.
4 J. Uglow, *The Lunar Men: the Friends who Made the Future*, London: Faber and Faber, 2002.
5 E. Robinson and D. McKie, *Partners in Science: James Watt and Joseph Black*, London: Constable, 1970.
6 S. Carnot, *Reflections on the Motive Power of Fire* [1824], R.H. Thurston and E. Mendoza (eds), New York, NY: Dover, 1960, pp. 3–5.
7 J. Gascoigne, *Science in the Service of Empire: Joseph Banks, the British State and the Uses of Science in the Age of Revolution*, Cambridge: Cambridge University Press, 1998.

8 D.M. Knight, *Humphry Davy: Science and Power*, 2nd edn, Cambridge: Cambridge University Press, 1998, p. 110.
9 H. Davy, *Collected Works*, London: Smith Elder, 1839–40, vol. 2, pp. 311–26.
10 J.Z. Fullmer, *Sir Humphry Davy's Published Works*, Cambridge, MA: Harvard University Press, 1969, pp. 70–3.
11 H. Davy, *Collected Works*, London: Smith Elder, 1839–40, vols 7 and 8.
12 T.R. Malthus, *First Essay on Population, 1798*, London: Macmillan, 1966.
13 H. Bence Jones, *The Life and Letters of Faraday*, 2nd edn, London: Longman, 1870, vol. 1, pp. v–vi.
14 *Life of William Allen*, London: Gilpin, 1846, vol. 1, pp. 33–5, 54, 150–1, 157.
15 G.E. Fussell, *Old English Farming Books, 1523–1793*, Aberdeen: Aberdeen Rare Books, 1978, part 2, pp. 70–151.
16 W. Marshall, *The Review and Abstract of the County Reports to the Board of Agriculture* [1808–15], Newton Abbot: David and Charles, 1968.
17 *Letters and Papers on Agriculture, Planting, &c. Selected from the Correspondence of the Bath and West of England Society*, 4th edn, 1 (1792), v–viii.
18 Ibid., 3rd edn, 2 (1792), v–viii.
19 M. Berman, *Social Change and Scientific Organization: the Royal Institution, 1799–1844*, London: Heinemann, 1978.
20 H. Davy, *Collected Works*, London: Smith Elder, 1839–40, vol. 6; Frank James' Presidential Address on the lamp will be published in the *Transactions of the Newcomen Society*.
21 H. Hellman, *Great Feuds in Technology: Ten of the Liveliest Disputes Ever*, Hoboken, NJ: Wiley, 2004, pp. 19–38.
22 W.R. Clanny, 'On the Means of Procuring a Steady Light in Coal Mines Without the Danger of Explosion', *Philosophical Transactions*, 103 (1813), 200–5.
23 J.B. Longmire, 'Remarks on the Wire-gauze Lamp Lately Constructed by Sir H. Davy', *Annals of Philosophy*, 8 (1816), 31–4. J.G. Children responded with what the editor (Thomas Thomson, of Edinburgh and Glasgow) called a panegyric of Davy, pp. 265–9; and J.H.H. Holmes, Longmire and others kept a controversy going, in a journal unsympathetic to the Royal Institution and to metropolitan science.
24 F.A.J.L. James discussed these researches of Davy in his 2004 Presidential Address to the Newcomen Society, which will be published in its *Transactions*.
25 D.M. Knight, *The Age of Science*, Oxford: Blackwell, 1992.
26 R.J. Richards, *The Romantic Conception of Life: Science and Philosophy in the Age of Goethe*, Chicago, IL: Chicago University Press, 2002.
27 C. Babbage, *Reflections on the Decline of Science in England, and on Some of its Causes*, London: Fellowes, 1830.
28 J. Morrell, *John Phillips and the Business of Victorian Science*, Aldershot: Ashgate, 2005, pp. 39–71, 101–32.
29 See F.A.J.L. James, 'An "Open Clash between Science and the Church"?', in D.M. Knight and M.D. Eddy (eds), *Science and Belief: from Natural Philosophy to Natural Science, 1700–1945*, Aldershot: Ashgate, 2005, pp. 171–93. On Wilberforce as moderniser and raconteur, see W. Callow, *An Autobiography*, H.M. Cundall (ed.), London: Black, 1908, p. 120.
30 D.S. Evans, T.J. Deeming, B.H. Evans and S. Goldfarb (eds), *Herschel at the Cape: Diaries and Correspondence of Sir John Herschel, 1834–1838*, Austin, TX: Texas University Press, 1969, p. 317.
31 A. Thwaite, *Glimpses of the Wonderful: the Life of Philip Henry Gosse*, London: Faber, 2002.
32 Royal Agricultural Society of England, *Handbook for the Week, with Diary of Appointments*, Southampton: Best and Snowden, 1844; N. Goddard, *Harvests of Change: the Royal Agricultural Society, 1838–1988*, London, 1988.
33 See the reviews by N. Fisher and S. Forgan in *BJHS*, 37 (2004), 478–80.

34 *Lectures on the Results of the Great Exhibition of 1851, Delivered before the Society of Arts, Manufactures and Commerce, at the Suggestion of HRH Prince Albert*, London: Bogue, 1852.
35 R. Bud and G.K. Roberts, *Science Versus Practice: Chemistry in Victorian Britain*, Manchester: Manchester University Press, 1986.
36 D. Edgerton, *Science, Technology and the British industrial 'decline', 1870–1970*, Cambridge: Cambridge University Press, 1996.
37 South Kensington Museum, *Conferences held in Connection with the Special Loan Collection of Scientific Apparatus: Physics and Mathematics*, and *Chemistry, Biology, Physical Geography, Geology, Mineralogy, and Meteorology*, and *Free Evening Lectures*, London: Chapman and Hall, 1876.
38 R. Anderson, '"What is Technology?": Education Through Museums in the Mid-19th Century', *BJHS*, 25 (1992), 169–84.
39 *Isaiah*, 7, 14; *Matthew*, 1, 23. See also A. Berlin and M.C. Brettler (eds), *The Jewish Study Bible*, Oxford: Oxford University Press, 2004, p. 798

7 Display

1 R. Whately, *Historic Doubts Relative to Napoleon Buonaparte*, 7th edn, London: Fellowes, 1841.
2 M.P. Crosland, *Science under Control*, Cambridge: Cambridge University Press, 1992.
3 J. Simon, *Pharmacy and the Chemical Revolution in France*, Aldershot: Ashgate, 2005.
4 J. Scott, *A Visit to Paris in 1814; with a view of the Moral, Political, Intellectual, and Social Condition of the French Capital*, 5th edn, pp. 230–318; and *Paris Revisited in 1815, by way of Brussels: including a Walk over the Field of Battle at Waterloo*, 4th edn, pp. 354–67, London: Longman, 1816, 1817.
5 Musée de Louvre, *Notice des Tableaux exposés dans la Galerie du Musée, des Statues, Bustes et Bas-reliefs, Suppléments, Explications*, Paris: Dubray, 1814.
6 J.P.F. Deleuze, *History and Description of the Royal Museum of Natural History*, Paris: Royer, 1825, pp. 117, 133, 256.
7 J. Morrell, *John Phillips and the Business of Victorian Science*, Aldershot: Ashgate, 2005, p. 19.
8 T.S. Raffles, *The History of Java* [1817], intr. J. Bastin, Oxford: Oxford University Press, 1978; S. Raffles, *Memoir of the Life and Public Services of Sir Thomas Stamford Raffles* [1830], intr. J. Bastin, Oxford: Oxford University Press, 1991, pp. 290, 590–5, 698–701.
9 T. Horsfield, *Zoological Researches in Java, and the Neighbouring Islands* [1821–4], intr. J. Bastin, Oxford: Oxford University Press, 1990.
10 G.C. Bompas, *Life of Frank Buckland*, new edn, London: Smith, Elder, 1887, pp. 46, 69, 100–3; F.T. Buckland, *Curiosities of Natural History*, 3rd edn, London: Bentley, 1858, pp. 297, 317.
11 V.M. Holt, *Why Not Eat Insects?* [1885], Faringdon; Classey, 1978.
12 N.A. Rupke, *Richard Owen: Victorian Naturalist*, New Haven, CT: Yale University Press, 1994.
13 J.W. Gruber and J.C. Thackray, *Richard Owen Commemoration: Three Studies*, London: Natural History Museum, 1992; R. Owen, *Hunterian Lectures in Comparative Anatomy, May–June 1837*, P.R. Sloan (ed.), London: Natural History Museum, 1992; N.A. Rupke, *Richard Owen: Victorian Naturalist*, New Haven, CT: Yale University Press, 1994.
14 W. Callow, *An Autobiography*, H.M. Cundall (ed.), London: Black, 1908, p. 136.
15 R. Owen, *Palaeontology: or a Systematic Summary of Extinct Animals and their Geological Relations* [1860], intr. D.M. Knight, London: Routledge/Natural History Museum, 2004.

16 A. Tennyson, *In Memoriam* [1850], S. Shatto and M. Shaw (eds), Oxford: Oxford University Press, 1982, p. 80.
17 R.J. Richards, *The Romantic Conception of Life*, Chicago, IL: Chicago University Press, 2002, pp. 517–18; the word goes back a long way: W. Pagel, 'Paracelsus and the Neoplatonic and Gnostic Tradition', *Ambix*, 8 (1960), 125–66, p. 159; also in A. Debus (ed.), *Alchemy and Early Modern Chemistry*, London: SHAC, 2005, pp. 101–42.
18 A. Desmond, *Archetypes and Ancestors: Palaeontology in Victorian London, 1850–1875*, London: Blond and Briggs, 1982.
19 W.H. Flower, *Essays on Museums and Other Subjects Connected with Natural History*, London: Macmillan, 1898.
20 R.W. Home (ed.), *Australian Science in the Making*, Cambridge: Cambridge University Press, 1988, chapters 3–6; B.W. Butcher, 'Darwin's Australian Correspondence; Deference and Collaboration in Colonial Science', in P.E. Rehbock (ed.), *Nature in its Greatest Extent: Western Science in the Pacific*, Honolulu: Hawaii University Press, 1988, pp. 139–57; Special issue, 'Colonial Science, *Isis*, 96 (2005), 52–87.
21 L. Pyenson and S. Sheets-Pyanson, *Servants of Nature: a History of Scientific Institutions, Enterprises and Sensibilities*, London: HarperCollins, 1999, pp. 125–49 and 150–72.
22 J. Browne, 'Biogeography and Empire', in N. Jardine, J.A. Secord and E.C. Spary (eds), *Cultures of Natural History*, Cambridge: Cambridge University Press, 1996, pp. 305–21.
23 E. O'Connor, *Raw Material: Producing Pathology in Victorian Culture*, Durham NC: Duke University Press, 2001.
24 W. Blunt, *In for a Penny; a Prospect of Kew Gardens*, London: Hamish Hamilton, 1978.
25 J. Gascoigne, *Science in the Service of Empire: Joseph Banks, the British State and the Uses of Science in the Age of Revolution*, Cambridge: Cambridge University Press, 1998.
26 T. Whittle and C. Cook, *Curtis's Flower Garden Displayed: 120 Plates from the Years 1787–1807*, Oxford: Oxford University Press, 1981.
27 N. Chambers (ed.), *The Letters of Sir Joseph Banks: a Selection, 1768–1820*, London: Imperial College Press, 2000, pp. 270–3.
28 L. Huxley (ed.), *The Life and Letters of Sir Joseph Dalton Hooker*, London: Murray, 1918, esp. vol. 2, pp. 159–77.
29 F. Burckhardt and S. Smith (eds), *The Correspondence of Charles Darwin*, Cambridge: Cambridge University Press, vol. 2, 1986, and subsequent volumes (in progress).
30 D.E. Allen, *Naturalists and Society: the Culture of Natural History in Britain, 1700–1900*, Aldershot: Ashgate Variorum, 2001, pp. V, 402–5; VI, 11–17; XIII, 206; J. Browne, 'Wardian Case', in J.L. Heilbron, *The Oxford Companion to the History of Modern Science*, Oxford: Oxford University Press, 2003, p. 826.
31 W. Blunt, *The Art of Botanical Illustration*, London: Collins, 1950, pp. 223–30, 268–82.
32 M. North, *A Vision of Eden*, intr. J.P.M. Brennan, A. Huxley, and B.E. Moon, Exeter: Webb and Bower, 1980.
33 South Kensington Museum, *Free Evening Lectures Delivered in Connection with the Special Loan Collection of Apparatus, 1876*, London: Chapman and Hall, 1876, pp. v–vi, 493, 519, 523–4.
34 J. Bennett, R. Brain, K. Bycraft, S. Schaffer, H.O. Sibum, and R. Staley, *Empires of Physics*, Cambridge: Whipple, 1993.
35 B. Marsden, 'The Progeny of those two "Fellows": Robert Willis, William Whewell and the Sciences of Mechanism, Mechanics and Machinery in Early Victorian Britain', *BJHS*, 37 (2004), 401–34.

36 R. Bud, S. Niziol, T. Boon and A. Nahum, *Inventing the Modern World: Technology Since 1750*, London: Science Museum, 2000.
37 S. Forgan and G. Gooday, 'Constructing South Kensington: the Buildings and Politics of T.H. Huxley's Working Environments', *BJHS*, 29 (1996), 435–68.
38 M.C.S. Christman, *1846: Portrait of a Nation*, Washington, DC: Smithsonian, 1996, pp. 191–8.
39 G. Pancaldi, 'Museums', in J.L. Heilbron (ed.), *The Oxford Companion to the History of Modern Science*, Oxford: Oxford University Press, 2003, pp. 550–1.
40 F.A.J.L. James, 'An "Open Clash Between Science and the Church"?: Wilberforce, Huxley and Hooker at the British Association, Oxford, 1860', in D.M. Knight and M.D. Eddy (eds), *Science and Beliefs*, Aldershot: Ashgate, 2005, pp. 171–93; on the museum complex, see J. Morrell, *John Phillips and the Business of Victorian Science*, Aldershot: Ashgate, 2005, pp. 307–27.
41 L.P. Williams, *Album of Science: the Nineteenth Century*, New York, NY: Scribner's, 1978, esp. pp. 360–81.
42 A.N. Wilson, *The Victorians*, London: Arrow, 2003, pp. 123–150.
43 R.M. Brain, 'Exhibitions', in J.L. Heilbron (ed.), *The Oxford Companion to the History of Modern Science*, Oxford: Oxford University Press, 2003, pp. 283–6.
44 R.M. Brain, 'Going to the Exhibition', in R. Staley (ed.), *The Physics of Empire*, Cambridge: Whipple, 1994.
45 J. Bennett, R.M. Brain, S. Schaffer, H.O. Sibum, and R. Staley, *1900: the New Age. A Guide to the Exhibition*, Cambridge: Whipple, 1994.
46 R.M. Brain, *Going to the Fair: Readings in the Culture of Nineteenth-century Exhibitions*, Cambridge: Whipple, 1993.
47 See the special issue of *Centaurus*, 39 (1997), 291–381.
48 J. Rüger, 'Nation, Empire and Navy: Identity Politics in the United Kingdom, 1887–1914', *Past and Present*, 185 (2004), 159–87, p. 184.
49 *Turbinia* is preserved in the Museum of Science and industry, Newcastle upon Tyne.
50 G.L. Geison, *The Private Science of Louis Pasteur*, Princeton, NJ: Princeton University Press, 1995, pp. 133, 267–9.
51 W.H. Helfand, *Quack, Quack, Quack: the Sellers of Nostrums in Prints, Posters, Ephemera and Books*, New York, NY: Grolier Club, 2002.
52 M. Fichman, *An Elusive Victorian: the Evolution of Alfred Russel Wallace*, Chicago, IL: Chicago University Press, 2004, p. 9.
53 J. Tyndall, *Fragments of Science*, 10th imp, London: Longman, 1899, vol. 2, pp. 453–90.
54 A. Desmond, *Huxley: Evolution's High Priest*, London: Michael Joseph, 1997, pp. 75–80.
55 C.C. Gillispie, *The Montgolfier Brothers and the Invention of Aviation, 1783–1784*, Princeton, NJ: Princeton University Press, 1983, pp. 118–20.
56 S. Winchester, *Krakatoa: the Day the World Exploded*, London: Penguin, 2004.

8 Travel

1 Some of the works referred to in this chapter are reprinted in facsimile in D.M. Knight (ed.), *Scientific Travellers*, London: Routledge and Natural History Museum, 2004.
2 N.J.W. Thrower, *The Three Voyages of Edmond Halley in the Paramore, 1698–1701*, London: Hakluyt Society, 1981.
3 For Halley's tortuous relations with John Flamsteed, his predecessor, see A. Johns, *The Nature of the Book: Print and Knowledge in the Making*, Chicago, IL: Chicago University Press, 1998, pp. 543–621.
4 W. St. Clair, *The Reading Nation in the Romantic Period*, Cambridge: Cambridge University Press, 2004, pp. 555–60.

5 F.A. Stafleu, *Linnaeus and the Linneans, the Spreading of their Ideas in Systematic Botany, 1735–89*, Utrecht: Oosthoek's Uitgeversmaatschappij, 1971.
6 J.C. Beaglehole, *The Life of Captain James Cook*, London: Hakluyt Society, 1974.
7 H. Carrington (ed.), *The Discovery of Tahiti: a Journal of the Second Voyage of HMS Dolphin round the world, by George Robertson, 1766–1768*, London: Halkuyt Society, 1948.
8 A. Dalrymple, *An Historical Collection of the Several Voyages and Discoveries in the South Pacific Ocean*, London: Nourse, 1770–1.
9 D. Howse, *Nevil Maskelyne, the Seaman's Astronomer*, Cambridge: Cambridge University Press, 1989.
10 A. Wood et al. (eds), *The Charts and Coastal Views of Captain Cook's Voyages: the Endeavour, 1768–71*, London: Hakluyt Society, 1988; J.C. Beaglehole (ed.), *The Journals of Captain James Cook*, 3 vols in 4 with atlas, Cambridge: Cambridge University Press, 1955–69.
11 P.A. Lanyon-Orgill, *Captain Cook's South Sea Island Vocabularies*, London: author, 1979.
12 D.J. Carr (ed.), *Sydney Parkinson: Artist of Cook's* Endeavour *Voyage*, London: Croom Helm, 1983; S. Parkinson, *A Journal of a Voyage to the South Seas*, London: S. Parkinson, 1773; P.J.P. Whitehead (ed.), *Forty Drawings of Fishes Made by the Artists who Accompanied Captain James Cook on his Three Voyages to the Pacific*, London: Natural History Museum, 1968.
13 D. Sobel, *Longitude*, London: Fourth Estate, 1996.
14 J. Cook, *A Voyage Towards the South Pole* [1777], Adelaide: Libraries Board of S. Australia, 1970; M.E. Hoare (ed.), *The* Resolution *Journal of Johann Reinhold Forster*, London: Hakluyt Society, 1982.
15 G. Forster, *Vögel der Südsee*, G. Steiner and L. Baege (eds), Leipzig: Insel-Verlag, 1971.
16 *The Voyage of Governor Phillip to Botany Bay*, London: Stockdale, 1789; E.H.J. Feeken, G.E.E. Feeken and O.H.K. Spate, *The Discovery and Exploration of Australia*, London: Nelson, 1970, is a good account of subsequent exploration, with many maps.
17 J. White, *Journal of a Voyage to New South Wales*, intr. R. Rienits, A.C. Chisholm (ed.), Sydney: Angus and Robertson, 1962; facsimile [1790] in D.M. Knight (ed.), *Scientific Travellers*, London: Routledge, 2004.
18 J. Hunter, *An Historical Journal of Events at Sydney and a Sea, 1787–1792*, J. Bach (ed.), Sydney: Angus and Robertson, 1968.
19 D. Collins, *An Account of the English Colony in New South Wales*, London: Cadell and Davies, 1798–1802; F. Schleiermacher, *On Religion: Speeches to its Cultured Despisers*, 2nd edn, R.Crouter (ed.), Cambridge: Cambridge University Press, 1996, p. xvii.
20 G. Dening, *Mr Bligh's Bad Language: Passion, Power and Theatre on the* Bounty, Cambridge: Cambridge University Press, 1992.
21 R. Cock, 'Scientific Servicemen in the Royal Navy and the Professionalisation of Science, 1816–55', in D.M. Knight and M.D. Eddy (eds), *Science and Beliefs: from Natural Philosophy to Natural Science, 1700–1900*, Aldershot: Ashgate, 2005, pp. 95–111.
22 M. Flinders, *A Voyage to Terra Australis: Undertaken for the Purpose of Completing the Discovery of that Vast Country*, London, Nichol, 1814.
23 C. Cornell (ed. and tr.), *Journal of Post Captain Nicolas Baudin, Commander in Chief of the Corvettes* Géographe *and* Naturaliste, Adelaide: Libraries Board of S. Australia, 1974, pp. 379–80.
24 J. Bonnemains, E. Forsyth, and B. Smith (eds), *Baudin in Australian Waters: the Artwork of the French Voyage of Discovery to the Southern Lands, 1800–1804*, Oxford: Oxford University Press, 1988.
25 A. Moyal, *A Bright and Savage Land: Scientists in Colonial Australia*, Sydney:

Collins, 1986; E. Lynn, *The Australian Landscape and its Artists*, Sydney: Bay Books, 1977; T. Bonyhady, *The Colonial Image: Australian Painting, 1800–1880*, Chippendale: Ellsyd Press, 1987.
26 J. Hackforth-Jones, *The Convict Artists*, Melbourne: Macmillan, 1977; R. McGowan, 'The Bank of England and the Policing of Forgery', *Past and Present*, 186 (2005), 81–116.
27 A. McEvey, *John Cotton's Birds of the Port Phillip District of New South Wales, 1843–1849*, Sydney: Collins, 1974.
28 B. Berzins, *The Coming of Strangers: Life in Australia, 1788–1822*, Sydney: State Library of NSW, 1988.
29 L.R. Hiatt and R. Jones, 'Aboriginal Conceptions of the Working of Nature', in R.W. Home (ed.), *Australian Science in the Making*, Cambridge: Cambridge University Press, 1988.
30 A. von Humboldt, *Personal Narrative of Travels to the Equinoctial Regions of America, During the Years 1799–1804*, tr. T. Ross, London: Bohn, 1852.
31 A. von Humboldt, *Aspects of Nature, in Different Lands and Different Climates: with Scientific Elucidations*, tr. E. Sabine, Longman and Murray, 1850.
32 D.B. Tyler, *The Wilkes Expedition: the First United States Exploring Expedition (1838–1841)*, Philadelphia, PA: American Philosophical Society, 1968.
33 R. Pineau (ed.), *The Japan Expedition 1852–1854: the Personal Journal of Commodore Matthew C. Perry*, intr. S.E. Morison, Washington, DC: Smithsonian, 1968.
34 A. Day, *The Admiralty Hydrographic Service 1795–1919*, London: HMSO, 1967.
35 G.L. Sullivan, *Dhow Chasing in Zanzibar Waters* [1873], London: Dawson, 1967; G.S. Graham, *Great Britain in the Indian Ocean, 1810–1850*, Oxford: Oxford University Press, 1967.
36 J.K. Tuckey, *Narrative of an Expedition to Explore the River Zaire, Usually called the Congo, in South Africa, in 1816*, J.Barrow (ed.), London: Murray, 1818.
37 W.F.W. Owen, *Narrative of Voyages to Explore the Shores of Africa, Arabia, and Madagascar*, London: Bentley, 1833: see vol. 1 pp. 231–2, 297–8, vol. 2, 224 on tars and science.
38 W. Ellis, *Three Visits to Madagascar During the Years 1853–1854–1856*, London: Murray, 1858.
39 W.H.B. Webster, *Narrative of a Voyage to the Southern Atlantic Ocean, in the Years 1828, 29, 30, Performed in H.M. Sloop* Chanticleer *under the Command of the late Captain Henry Foster* [1834], Folkestone: Dawson, 1970; R. Cock, 'Scientific Servicemen in the Royal Navy and the Professionalisation of Science, 1816–55, in D.M. Knight and M.D. Eddy (eds), *Science and Beliefs*, Aldershot: Ashgate, 2005, p. 105.
40 R.D. Keynes (ed.), *The Beagle Record*, Cambridge: Cambridge University Press, 1979; J. Browne, *Charles Darwin: Voyaging*, London: Cape, 1995.
41 J. Huxley (ed.), *T.H. Huxley's Diary of the Voyage of HMS* Rattlesnake, New York, NY: Doubleday, 1936; P. White, *Thomas Huxley: Making the 'Man of Science'*, Cambridge: Cambridge University Press, 2003; A. Desmond, *Huxley*, London: Michael Joseph, 1994–7.
42 F. Burkhardt *et al.* (eds), *The Correspondence of Charles Darwin*, vol. 12, Cambridge: Cambridge University Press, 2001, p. 395.
43 J.C. Ross, *A Voyage of Discovery and Research in the Southern and Antarctic Regions, During the Years 1839–43* [1847], Newton Abbot: David & Charles, 1969.
44 S. Ruskin, *John Herschel's Cape Voyage: Private Science, Public Imagination, and the Ambitions of Empire*, Aldershot: Ashgate, 2004.
45 J.L. Stokes, *Discoveries in Australia with an Account of the Coasts and Rivers Explored and Surveyed During the Voyage of HMS* Beagle *in the Years 1837–38–39–40–41–42–43*, London: Boone, 1846, vol. 2, p. 6.
46 P.P. King, *Narrative of a Survey Voyage of the Intertropical and Western Coasts of Australia, Performed Between the Years 1818 and 1822*, London: Murray, 1827.

47 J. Buzzard, *The Beaten Track: European Tourism, Literature, and the Ways to Culture, 1800–1918*, Oxford: Oxford University Press, 1993.
48 T.H. Levere and R.A. Jarrell (eds), *A Curious Field-book: Science and Society in Canadian History*, Toronto: Oxford University Press, 1974.
49 J. Franklin, *Narrative of a Journey to the Shores of the Polar Sea in the Years 1819, 20, 21, & 22* [1825], Rutland VT: Tuttle, 1970; C.S. Houston (ed.), *To the Arctic by Canoe: the Journal and Paintings of Robert Hood, Midshipman with Franklin*, Montreal: McGill-Queen's University Press, 1974; C.S. Houston (ed.), *Arctic Ordeal: the Journal of John Richardson, Surgeon-Naturalist with Franklin, 1820–22*, Montreal: McGill-Queen's University Press, 1984.
50 T. Hearne, *Journey from Prince of Wales's Fort in Hudson's Bay to the Northern Ocean*, London: Cadell, 1795; W.K. Lamb (ed.), *The Journals and Letters of Sir Alexander Mackenzie*, Cambridge: Hakluyt Society, 1970.
51 G. Vancouver, *A Voyage of Discovery to the North Pacific Ocean, and Round the World*, London: Robinson, 1798; W.K. Lamb (ed.), *The Voyage of George Vancouver, 1791–1795*, London: Hakluyt Society, 1984.
52 J. Franklin, *Narrative of a Second Expedition to the Shores of the Polar Sea, in the Years 1825, 1826, and 1827*, London: Murray, 1828; F.W. Beechey, *Narrative of a Voyage to the Pacific and Behring's Strait*, London: Colburn and Bentley, 1831.
53 F.L. McClintock, *A Narrative of the Discovery of the Fate of Sir John Franklin and his Companions*, London: Murray, 1859; W. Barr (ed.), *Searching for Franklin: the Land Arctic Searching Expedition, 1855*, London: Hakluyt Society, 1999.
54 J.F.W. Herschel (ed.), *A Manual of Scientific Enquiry; Prepared for the Use of the Officers of Her Majesty's Navy; and Travellers in General* [1849], intr. D.M. Knight, Folkestone: Dawson, 1974.
55 M. Faraday, *Chemical Manipulation* [1827], in D.M. Knight (ed.), *The Development of Chemistry*, London: Routledge, 1998, vol. 5.
56 E.V. Brunton, *The* Challenger *Expedition, 1872–1876: a Visual Index*, 2nd edn, London: Natural History Museum, 2005; G.S. Ritchie, *The Admiralty Chart: British Naval Hydrography in the Nineteenth Century*, London: Hollis and Carter, 1967, pp. 323–7, 330–7.
57 C.W. Thomson, *The Voyage of the* Challenger*: the Atlantic*, London: 1877; R. Corfield, *The Silent Landscape: in the Wake of HMS* Challenger *1872–1876*, London: Murray, 2004.
58 E. Wilson, *Diary of the* Discovery *Expedition to the Antarctic, 1901–1904*, A. Savours (ed.), London: Blandford, 1966; *Birds of the Antarctic*, B. Roberts (ed.), London: Blandford, 1967.
59 H.W. Bates, *The Naturalist on the River Amazons*, intr. E. Clodd, London: Murray, 1892; the first edition is reprinted in D.M. Knight (ed.), *Scientific Travellers*, London: Routledge, 2004; F. Burckhardt *et al.*, *The Correspondence of Charles Darwin*, Cambridge: Cambridge University Press, 1985–, vols 9, p. 92; 10, pp. 5–6, 11, pp. 323, 326, 330; J. Dickinson, 'Getting In on his Rambles in South America: the Published Correspondence of H.W. Bates in the Amazon, 1848–59', *Archives of Natural History*, 23 (1996), 201–8.
60 T. Forrest, *A Voyage to New Guinea and the Moluccas, 1774–1776* [1780], intr. D.K. Bassett, Oxford: Oxford University Press, 1969.
61 W. Marsden, *The History of Sumatra* [1811], intr. J. Bastin: Oxford: Oxford University Press, 1966; see also J. Anderson, *Mission to the East Coast of Sumatra in 1823* [1826], intr. N. Tarling, Oxford: Oxford University Press, 1971.
62 J. Crawfurd, *Journal of an Embassy to the Courts of Siam and Cochin China* [1828], Oxford: Oxford University Press, 1967; *A Descriptive Dictionary of the Indian Islands and Adjacent Countries* [1856], Oxford: Oxford University Press, 1971.
63 See e.g. M. Bentley, *Lord Salisbury's World: Conservative Environments in Late-Victorian Britain*, Cambridge: Cambridge University Press, 2001, pp. 221–5.

64 A.R. Wallace, *The Malay Archipelago* [1869], in D.M. Knight (ed.), *Scientific Travellers*, London: Routledge, 2004.
65 A.R. Wallace, *Darwinism: an Exposition of the Theory of Natural Selection with Some of its Applications*, London: Macmillan, 1889.
66 M. Fichman, *An Elusive Victorian: the Evolution of Alfred Russel Wallace*, Chicago, IL: Chicago University Press, 2004.
67 F. Darwin and A.C. Seward (eds), *More Letters of Charles Darwin*, London: Murray, 1903, vol. 2, p. 360.
68 T. Belt, *The Naturalist in Nicaragua*, London: Murray, 1874, p. 196.
69 W.H. Hudson, *The Naturalist in La Plata*, London: Chapman and Hall, 1892.
70 W.H. Hudson, *Birds in London* [1898], Newton Abbot: David and Charles, 1969; *The Book of a Naturalist*, London: Hodder and Stoughton, n.d.; *A Shepherd's Life* [1910], Tisbury: Compton Press, 1978.
71 I. Jenkins and K. Sloan, *Vases and Volcanoes: Sir William Hamilton and his Collection*, London: British Museum, 1996.
72 R. Chandler, *Travels in Asia Minor, 1764–1765*, ed. and abr. E. Clay, intr. A. Wilton, London: British Museum, 1971; C. Foss, *Ephesus after Antiquity*, Cambridge: Cambridge University Press, 1979, pp. 176–8.
73 J. Gascoigne, *Science in the Service of Empire: Joseph Banks, the British State and the Uses of Science in the Age of Revolution*, Cambridge: Cambridge University Press, 1998, pp. 179–82.
74 A. Smith, *Illustrations of the Zoology of South Africa*, London: Smith Elder, 1839–40: the *Annulosa* (invertebrates), W.S. McLeay (ed.) were published in 1838.
75 W. Monk (ed.), *Dr. Livingstone's Cambridge Lectures*, Cambridge: Deighton and Bell, 1858; T. Hughes, *Life of Dr Livingstone*, London: Macmillan, 1889, was in its eleventh reprinting when given to my father as a school prize in 1916; R. Foskett (ed.), *The Zambesi Doctors: David Livingstone's Letters to John Kirk, 1858–1872*, Edinburgh: Edinburgh University Press, 1964.
76 D. Middleton (ed.), *The Diary of A.J. Mounteney Jephson: the Emin Pasha Relief Expedition 1887–1889*, Cambridge: Hakluyt Society, 1969.
77 W.C. Harris, *Portraits of the Game and Wild Animals of Southern Africa*, intr. A.C. Tabler and R. Liversidge, Cape Town: Balkema, 1969.
78 W. Moorcroft, *Travels in the Himalayan Provinces of Hindustan and the Panjab* [1841], intr. G.J. Alder, Oxford: Oxford University Press, 1979.
79 J. Lawrence and A. Woodwiss, *The Journals of Honoria Lawrence: India Observed 1837–1854*, London: Hodder and Stoughton, 1980.
80 M. North, *A Vision of Eden*, intr. J.M.P. Brennan, A. Huxley and B.E. Moon, Exeter: Webb and Bower, 1980; this is an illustrated abridgement of *Recollections of a Happy Life*, London: Macmillan, 1892.
81 F. Galton: *The Art of Travel: or, Shifts and Contrivances Available in Wild Countries*, London: Murray, 1855 – 8th edn, 1893.

9 Imagining

1 L. Wolpert, *The Unnatural Nature of Science*, London: Faber, 1992.
2 H. Davy, *Consolations in Travel: or, the Last Days of a Philosopher*, London: Murray, 1830, pp. 248–50.
3 S.T. Coleridge, *The Friend*, B.E. Rooke (ed.), London: Routledge, 1969, vol. 1, pp. 470–1.
4 J. Smith, *Fact and Feeling: Baconian Science and the Nineteenth-century Literary Imagination*, Madison: Wisconsin University Press, 1994.
5 J.L. Lowes, *The Road to Xanadu; a Study in the Ways of the Imagination*, Boston: Houghton Mifflin, new edn, 1955; R. Holmes, *Coleridge: Early Visions*, London:

Hodder and Stoughton, 1989, pp. 169–204; R. Holmes (ed.), *Coleridge: Selected Poems*, London: HarperCollins, 1996, pp. 81–100.
6. W. Wordsworth, *Poetry and Prose*, W.M. Merchant (ed.), London: Hart-Davis, 1955, pp. 228, 230, 247; W. Shakespeare, *Midsummer-Night's Dream*, V, 1, line 8.
7. J.F.W. Herschel, *A Preliminary Discourse on the Study of Natural Philosophy* [1830], intr. M. Partridge, New York, NY: Johnson, 1966.
8. T. Nagel (ed.), *J.S. Mill's Philosophy of Scientific Method*, New York, NY: Hafner, 1950.
9. M. Fisch and S. Schaffer (eds), *William Whewell: a Composite Portrait*, Oxford: Oxford University Press, 1991; M. Fisch, *William Whewell: Philosopher of Science*, Oxford: Oxford University Press, 1991; William Whewell, *History of the Inductive Sciences*, 3rd edn, London: Parker, 1857; *Philosophy of the Inductive Sciences*, 2nd edn, London: Parker, 1847.
10. A. Comte, *The Positive Philosophy*, ed. and tr. H. Martineau, London: Chapman, 1853, pp. 1–2.
11. T.R. Wright, *The Religion of Humanity: the Impact of Comtean Positivism on Victorian Britain*, Cambridge: Cambridge University Press, 1986; T. Dixon, 'The Invention of Altruism: Auguste Comte's *Positive Polity* and Respectable Unbelief in Victorian Britain', in D.M. Knight and M.D. Eddy (eds), *Science and Beliefs*, Aldershot: Ashgate, 2005, pp. 195–211.
12. A. Comte, *The Catechism of Positive Religion*, 3rd edn, tr. R. Congreve, London: Kegan Paul, 1891.
13. J.R. Watson (ed.), *Everyman's Book of Victorian Verse*, London: Dent, 1982, pp. 240–1.
14. H. Helmholtz, *Popular Lectures on Scientific Subjects*, trans. E. Atkinson *et al.*, intr. J. Tyndall, London: Longman, 1873.
15. D.M. Knight: *Science and Spirituality: the Volatile Connection*, London: Routledge, 2004, pp. 82–91.
16. J. Tyndall, *The Glaciers of the Alps; Mountaineering in 1861*, intr. J. Lubbock, London: Dent, 1906, pp. 239–40.
17. W. Vaughan, *Friedrich*, London: Phaidon, 2004.
18. J. Tyndall, *Heat a Mode of Motion*, 10th edn, London: Longman, 1894.
19. W. Shakespeare, *The Tempest*, I, ii, 50.
20. J. Tyndall, *The Forms of Water in Clouds & Rivers, Ice & Glaciers*, 4th edn, London: King, 1874, p. x.
21. J. Tyndall, *Fragments of Science*, 6th edn, London: Longman, 1899, vol. 2, pp. 101–134; quotations are from pp. 104, 124, 127, 131, 132, 134.
22. H. Mansel, *The Limits of Religious Thought Examined in Eight Lectures*, 4th edn, London: Murray, 1859.
23. H. Davy, *Collected Works*, London: Smith Elder, 1839–40, vol. 7, pp. 93–9.
24. D.M. Knight, *Atoms and Elements*, 2nd edn, London: Hutchinson, 1970, and *Ideas in Chemistry: a History of the Science*, London: Athlone, 1992, pp. 116–27; A.J. Rocke, *Chemical Atomism in the Nineteenth Century*, Columbus, OH: Ohio University Press, 1984.
25. W. Thomson, Baron Kelvin, 'The Size of Atoms', *Popular Lectures and Addresses*, 2nd edn, London, 1892, vol. 1, pp. 154–224.
26. D. Cahan (ed.), *Hermann von Helmholtz and the Foundations of Nineteenth-century Science*, Berkeley, CA: California University Press, 1993.
27. G. Gigerenzer, Z. Swijtink, T. Porter, L. Dalston, J. Beatty and L. Kruger, *The Empire of Chance: How Probability Changed Science and Everyday Life*, Cambridge: Cambridge University Press, 1989.
28. B. Mahon, *The Man Who Changed Everything: the Life of James Clerk Maxwell*, Chichester: Wiley, 2003, p. 82; J.C. Maxwell, *Scientific Papers*, W.D. Niven (ed.), Cambridge; Cambridge University Press, 1890, vol. 1, pp. 377–409, vol. 2 pp. 1–78, 361–77.

29 W. Thomson, Baron Kelvin, 'The Sorting Demon of Maxwell', *Popular Lectures and Addresses*, 2nd edn, London, 1892, vol. 1, pp. 144–8.
30 F. Burkhardt *et al.* (eds), *The Correspondence of Charles Darwin*, vol. 9, Cambridge: Cambridge University Press, 1994, pp. 135–6.
31 J.D. Burchfield, *Lord Kelvin and the Age of the Earth*, New York, NY: Science History, 1975.
32 H.G. Wells, *The Time Machine* [1895], J. Lawton (ed.), London: Everyman, 1995.
33 W. Thomson, Baron Kelvin, *Popular Lectures and Addresses*, London: Macmillan, 1894, vol. 3, pp. 10–64.
34 F.A.J.L. James 'An "Open Clash Between Science and the Church"?', in D.M. Knight and M.D. Eddy (eds), *Science and Beliefs*, Aldershot: Ashgate, 2005, pp. 171–94.
35 'Reviews', *Quarterly Journal of Science*, 1 (1864), 544.
36 R. Somerset, 'Darwinian "Becoming" and Early 19th-century Historiography: the Cases of Jules Michelet and Thomas Carlyle', in D.M. Knight and M.D. Eddy (eds), *Science and Beliefs*, Aldershot: Ashgate, 2005, pp. 129–39.
37 W.H. Brock, *Science for All: Studies in the History of Victorian Science and Education*, Aldershot: Ashgate Variorum, 1996; D. Edgerton, *Science, Technology, and the British Industrial 'Decline', 1870–1970*, Cambridge: Cambridge University Press, 1996.

10 Science gossip

1 G. Cantor, G. Dawson, G. Gooday, R. Noakes, S. Shuttleworth and J.R. Topham, *Science in the Nineteenth-century Periodical: Reading the Magazine of Nature*, Cambridge: Cambridge University Press, 2004; G. Cantor and S. Shuttleworth (eds), *Science Serialized: Representations of the Sciences in Nineteenth-century Periodicals*, Boston, MA: MIT; L. Henson, G. Cantor, G. Dawson, R. Noakes, S. Shuttleworth and J.R. Topham (eds), *Culture and Science in the Nineteenth-century Media*, Aldershot: Ashgate, 2004. My essay-review of these, 'Snippets of Science', will appear in *Studies in the History and Philosophy of Science*. See also V. Gray, *Charles Knight: Educator, Publisher, Writer*, Aldershot: Ashgate, 2006.
2 See e.g. D. May, *Critical Times: the History of the* Times Literary Supplement, London: HarperCollins, 2001.
3 P. Metcalf, *James Knowles: Victorian Editor and Architect*, Oxford: Oxford University Press, 1980, pp. 274–351.
4 B. Lightman, 'Fighting Even with Death: Balfour, Scientific Naturalism, and Thomas Henry Huxley's Final Battle', in A.P. Barr (ed.), *Thomas Henry Huxley's Place in Science and Letters*, Athens, GA: Georgia University Press, 1997, pp. 323–50.
5 A. de Morgan, *A Budget of Paradoxes* [1872, 2nd edn, 1915], facsimile reprint, New York: Books for Libraries, 1969, vol. 1, pp. 15–21.
6 E.g. J.F.W. Herschel, *Essays from the Edinburgh and Quarterly Reviews, with Addresses and Other Pieces*, London: Longman, 1857; S. Wilberforce, *Essays Contributed to the Quarterly Review*, London: Murray, 1874.
7 A. de Morgan, *A Budget of Paradoxes* [1915], Freeport, NY: Books for Libraries, 1969, vol. 1, pp. 15–21.
8 In *Macmillan's Magazine* 51 (1884–5) we find a mixture of the signed, pseudonymous and anonymous.
9 *Essayes of Natural Experiments made in the Academie del Cimento, under the Protection of the Most Serene Prince Leopold of Tuscany, Written in Italian by the Secretary of that Academy*, tr. R. Waller [1684], intr. A.R. Hall, New York, NY: Johnson, 1964
10 *Abstracts of the Papers Printed in the Philosophical Transactions*, 1 (1832).
11 *Proceedings of the Royal Society of London (Being a Continuation of Abstracts . . .)* 7 (1856).

12 R.B. Williams, 'Three Unrecorded Publications by Philip Henry Gosse, with Notes on Primary Text on Wrappers of Journals and Books, and a Proposal for the Bibliographical Description of Wrappers', *Archives of Natural History*, 32 (2005), 34–40.
13 D.M. Knight, 'The Case of *Annals of Science*', in M. Beretta, C. Pogliano and P. Redondi (eds), *Journals and the History of Science*, Florence: Olschki, 1999, pp. 153–66.
14 I happen to have twelve offprints of papers by the botanist and horticulturist Thomas Andrew Knight, from *Philosophical Transactions* for years between 1807 and 1837.
15 There were twenty shillings to a pound, and twelve pence to a shilling, in the old system; and in comparing prices with today's, we should probably multiply by 50 or 60: J. Morrell, *John Phillips and the Business of Victorian Science*, Aldershot: Ashgate, 2005, p. xvi.
16 K. Hufbauer, *The Formation of the German Chemical Community (1720–1795)*, Berkeley, CA: California University Press, 1982.
17 M.P. Crosland, *In the Shadow of Lavoisier: the* Annales de Chimie, *and the Establishment of a New Science*, Chalfont St Giles: BSHS, 1994.
18 W.H. Brock and A.J. Meadows, *The Lamp of Learning: Taylor and Francis and the Development of Science Publishing*, London: Taylor and Francis, 1984.
19 J.Z. Buchwald and A. Warwick (eds), *Histories of the Electron: the Birth of Microphysics*, Cambridge, MA: MIT Press, 2004.
20 D.M. Knight, *The Transcendental Part of Chemistry*, Folkestone: Dawson, 1978, pp. 155–84; papers relevant to this are reprinted in facsimile in D.M. Knight (ed.), *Classical Scientific Papers: Chemistry*, London: Mills and Boon, 2 vols, 1968–70.
21 R. Fox and A. Nieto-Galan (eds), *Natural Dyestuffs and Industrial Culture in Europe, 1750–1880*, Canton, MA: Science History, 1999.
22 L. Newlyn (ed.), *The Cambridge Companion to Coleridge*, Cambridge: Cambridge University Press, 2002, p. 129.
23 *Register of the Arts and Sciences, Containing a Correct Account of Several Hundred of the Most Important and Interesting Inventions, Discoveries, and Processes*, 1 (1824), 177, 212; see also 75, 129, 287 and 2 (1825), 225, 261ff, 353.
24 C. Burney and W. Harris (eds), *A Catalogue of the Library of the Royal Institution of Great Britain*, 2nd edn, London: Royal Institution, 1821.
25 *The Quarterly Journal of Science, Literature, and the Arts*, 8 (1820), 234–40; quotation from p. 234. See also 10 (1821) 215–16, 217; 11 (1821), 240–6; 18 (1825), 117–35, 199–200, 201–23.
26 A. Hayter, *Opium and the Romantic Imagination*, London: Faber, 1968.
27 *The Chemist*, 1 (1824).
28 *The Mechanics Magazine*, 1 (1823)–97 (1873).
29 *Magazine of Natural History, and Journal of Zoology, Botany, Mineralogy, Geology, and Meteorology*, 2 (1829), iii–iv.
30 *The Magazine of Natural History*, 3 (1830), 297–308; 389–90; the Swainson controversy, 4 (1831), 97–108, 199–207, 316–337, 455–9, 481–8, 559–60; 5 (1832), [109–12], [191–208].
31 T. Horsfield, *Zoological Researches in Java, and the Neighbouring Islands* [1824], intr. J. Bastin, Oxford: Oxford University Press, 1990, unpaginated but see plate of Tapirus Malayanus, and description.
32 *Magazine of Natural History*, 6 (1833), 464–8; 7 (1834), 66–72; 1–6; 9 (1836), 130–8, 175–82.
33 *The Gentleman's Magazine*, 102 (1832), 9, 73, 558.
34 *The Quarterly Journal of Science*, 1 (1864), 1–2.
35 D.M. Knight, 'Science and Culture in Mid-Victorian Britain: the Reviews, and William Crookes' *Quarterly Journal of Science*', *Nuncius*, 11 (1996), 43–54.
36 *The Quarterly Journal of Science*, 1 (1864), 545–54, 2 (1865), 187–98.
37 *The Quarterly Journal of Science*, 8 (1871), 'War Science', 43–59, 520–6, 'Royal

College of Chemistry', 145–53, 'Great Pyramid', 16–35, 177–214, 'Psychic Force', 339–49, 471–93, 'Gun-cotton', 494–503, 'Patent rights', 504–20.
38 *The Quarterly Journal of Science*, 8 (1871), 144.
39 *The Journal of Science*, 16 (1879), i.
40 W.H. Brock, 'The Making of an Editor: the case of William Crookes', L. Henson *et al.* (eds), *Culture and Science in the Nineteenth-century Media*, Aldershot: Ashgate, 2004, pp. 189–98.
41 B. Lightman, '*Knowledge* confronts *Nature*: Richard Proctor and Popular Science Periodicals', P.C. Kjærgaard, '"Within the Bounds of Science": Redirecting Controversies to *Nature*', and R. Barton, 'Scientific Authority and Scientific Authority in *Nature*: North Britain against the X-club', L. Henson *et al.* (eds), *Culture and Science in the Nineteenth-century Media*, Aldershot: Ashgate, 2004, pp. 199–210, 211–21, 223–35.
42 *Science-Gossip*, new series 1, no. 1 (February 1894), 1–2.
43 *Science-Gossip*, 5, no. 53 (October 1898), 154.
44 *Science-Gossip*, 8, no. 93 (February 1902), 275.

11 Suspending judgement

1 On Huxley, see the two-volume set, A.P. Barr (ed.), *The Major Prose of Thomas Henry Huxley*, and *Thomas Henry Huxley's Place in Science and Letters: Centenary Essays*, Athens, GA: Georgia University Press, 1997; R.A. Jarrell, 'Visionary or Bureaucrat? T.H. Huxley, the Science and Art Department and Science Teaching for the Working Class', *Annals of Science*, 55 (1998), 219–40.
2 F.N. Egerton, *Hewett Cottrell Watson: Victorian Plant Ecologist and Evolutionist*, Aldershot: Ashgate, 2003.
3 A. Desmond, *Huxley*, 2 vols, London: Michael Joseph, 1994–7; P. White, *Thomas Huxley: Making the 'Man of Science'*, Cambridge: Cambridge University Press, 2003.
4 J. Butler, *The Analogy of Religion, Natural and Revealed*, new edn, London: Rivington, 1791, p. 3.
5 R. Descartes, *A Discourse of a Method For the Well Guiding of Reason, and the Discovery of Truth in the Sciences*, London: Newcombe, 1649.
6 W. Henry, *Philosophical Magazine*, 7 (1830), 229; J.A. Paris, *The Life of Sir Humphry Davy*, London: Colburn and Bentley, 1831, pp. 96–8.
7 L. Newlyn (ed.), *The Cambridge Companion to Coleridge*, Cambridge: Cambridge University Press, 2002, p. 115.
8 S.T. Coleridge, *The Friend* [1818], B.E. Rooke (ed.), London: Routledge, 1969, vol. 1, pp. 448–524; 463, 482.
9 S.T. Coleridge, *Hints Towards the Formation of a More Comprehensive Theory of Life*, S.B. Watson (ed.), London: Churchill, 1848, p. 32.
10 W. Lawrence, *Lectures on Physiology, Zoology, and the Natural History of Man, Delivered at the Royal College of Surgeons*, new edn, London: Benbow, 1822.
11 T. Fulford (ed.), *Romanticism and Science, 1773–1833*, London: Routledge, 2002, vol. 5, pp. 45–83, reprints documents by Abernethy and Lawrence.
12 D.M. Knight, 'From Science to Wisdom: Humphry Davy's Life', in M. Shortland and R. Yeo (eds), *Telling Lives: Essays on Scientific Biography*, Cambridge: Cambridge University Press, 1996, pp. 103–114.
13 J.F.W. Herschel, *A Preliminary Discourse on the Study of Natural Philosophy* [1830], intr. M. Partridge, New York, NY: Johnson, 1966; D.G. King-Hele (ed.), *John Herschel 1792–1871: a Centenary Commemoration*, London: Royal Society, 1992.
14 S. Ruskin, *John Herschel's Cape Voyage: Private Science, Public Imagination and the Ambitions of Empire*, Aldershot: Ashgate, 2004.
15 M. Fisch and S. Schaffer (eds), *William Whewell: a Composite Portrait*, Oxford:

Oxford University Press, 1991; R. Yeo, *Defining Science: William Whewell, Natural Knowledge and Public Debate in Early Victorian Britain*, Cambridge: Cambridge University Press, 1993.
16. R. Cudworth, *The True Intellectual System of the Universe*, London: Royston, 1678, p. 888.
17. G.G. Stokes, *Mathematical and Physical Papers*, vol. 2, Cambridge: Cambridge University Press, 1883, p. 97.
18. R. Yeo, *Defining Science: William Whewell, Natural Knowledge, and Public Debate in Early Victorian Britain*, Cambridge: Cambridge University Press, 1993.
19. F. Burkhardt et al., *The Correspondence of Charles Darwin*, vol. 9, 1994, pp. xv, 204.
20. P. Corsi, *Science and Religion; Baden Powell and the Anglican Debate, 1800–1860*, Cambridge: Cambridge University Press, 1988; B. Powell, *The Connexion of Natural and Divine Truth; or, the Study of the Inductive Philosophy Considered as Subservient to Theology*, London: Parker, 1858.
21. B. Powell, *Essays on the Spirit of the Inductive Philosophy, the Unity of Worlds, and the Philosophy of Creation*, London: Longman, 1855.
22. B. Jowett et al., *Essays and Reviews*, London: Parker, 1860.
23. B. Powell, *Essays*, 1855, pp. 441, 36.
24. H.C. Oersted, *The Soul in Nature*, London: Bohn, 1852.
25. H.C. Oersted, *Selected Scientific Works*, tr. and ed. K. Jelved, A.D. Jackson, O. Knudsen and A.D. Wilson, Princeton, NJ: Princeton University Press, 1998. There was a conference at Harvard in 2002 on Oersted, which will be published.
26. H.C. Oersted, *The Soul in Nature*, London: Bohn, 1852, pp. xviii, 183.
27. W. Crookes, 'Spiritualism', *Quarterly Journal of Science*, 7 (1870), 317.
28. F. Burkhardt et al. (eds), *The Correspondence of Charles Darwin*, vol. 4, Cambridge: Cambridge University Press, 1988, p. 488.
29. L. Oken, *Elements of Physiophilosophy*, tr. A. Tulk, London: Ray Society, 1847, pp. 1, 270.
30. R.J. Richards, *The Romantic Conception of Life: Science and Philosophy in the Age of Goethe*, Chicago, IL: Chicago University Press, 2002, esp. pp. 497–502.
31. W. Goethe, *Theory of Colours*, tr. C.L. Eastlake, London: Murray, 1840.
32. J. Hamilton (ed.), *Fields of Influence: Conjunctions of Artists and Scientists, 1850–1860*, Birmingham: Birmingham University Press, 2001; Hamilton curated the exhibition 'Turner and the Scientists' at the Tate Gallery in London in 2000.
33. M.A. di Gregorio, 'Thomas Henry Huxley and German Science', in A.P. Barr (ed.), *Thomas Henry Huxley's Place in Science and Letters: Centenary Essays*, Athens, GA: Georgia University Press, 1997, pp. 159–81.
34. P. White, *Thomas Huxley: Making the 'Man of Science'*, Cambridge: Cambridge University Press, 2003.
35. *Proverbs*, 9, 10.
36. B. Lightman, *The Origins of Agnosticism: Victorian Unbelief and the Limits of Knowledge*, Baltimore, CT: Johns Hopkins University Press, 1987, pp. 32–67; H. Mansel, *The Limits of Religious Thought Examined*, 4th edn, London: Murray, 1859.
37. T.R. Wright, *The Religion of Humanity: the Impact of Comtean Positivism on Victorian Britain*, Cambridge: Cambridge University Press, 1986; A. Comte, *The Catechism of Positive Religion*, 3rd edn, tr. R. Congreve, London: Kegan Paul, Trench, Trübner, 1891.
38. A. Comte, *The Positive Philosophy*, tr. H. Martineau, London: Chapman, 1853.
39. D.M. Knight, 'Huxley and Philosophy of Science', in A.P. Barr, *Thomas Henry Huxley's Place in Science and Letters*, Athens, GA: Georgia University Press, 1997, pp. 51–66.
40. T.H. Huxley, *Hume*, London: Macmillan, 1878.
41. O. Lodge (ed.), *Huxley Memorial Lectures*, Birmingham: Cornish, 1914, p. 36.
42. *Royal Commission on Scientific Instruction and the Advancement of Science*, London: Eyre and Spottiswoode for HMSO, 1872–3–5, vol. 1, pp. xx, 628; vol. 2, pp. lvi–lx.

43 J. Morrell, *John Phillips and the Business of Victorian Science*, Aldershot: Ashgate, 2005, pp. 310–24.
44 M. Travers, *A Life of Sir William Ramsay*, London: Edward Arnold, 1956, pp. 72–80.
45 *Royal Commission on Scientific Instruction* ..., vol. 2, London: HMSO, 1873, 4th report, p. 9.

12 Classical physics

1 R. Yeo, 'Encyclopaedic Knowledge', in M. Frasca-Spada and N. Jardine (eds), *Books and the Sciences in History*, Cambridge: Cambridge University Press, 2000, 207–24, p. 218 for 1741; and see T. Young, *A Course of Lectures on Natural Philosophy and the Mechanical Arts*, London: Johnson, 1807.
2 T. Dixon, 'The Invention of Altruism: Auguste Comte's *Positive Polity* and Respectable Unbelief in Victorian Britain', in D.M. Knight and M.D. Eddy (eds), *Science and Belief: from Natural Philosophy to Natural Science, 1700–1900*, Aldershot: Ashgate, 2005, pp. 195–211, see p. 200.
3 C. Smith and M.N. Wise, *Energy and Empire: a Biographical Study of Lord Kelvin*, Cambridge: Cambridge University Press, 1989, pp. 149–494.
4 S. Carnot, *Réflexions sur la Puissance Motrice du Feu* [1824], R. Fox (ed.), Paris: Vrin, 1978; *Reflections on the Motive Power of Fire; and Other Papers on the Second Law of Thermodynamics by É.Clapeyron and R.Clausius*, E. Mendoza (ed.), New York: Dover, 1960.
5 W. Shakespeare, *King Lear*, 1, 1, 92.
6 A conference on Tom Wedgwood, held at the Royal Institution on 12 May 2005, had a catalogued exhibition: A.S. Barnes (ed.), *Tom Wedgwood (1771–1805): a Lunar Son and his Influences*, London: Royal Institution, 2005.
7 H.C. Ørsted, *Selected Scientific Works*, ed. and tr. K. Jelved, A.D. Jackson. O. Knudsen and A.D. Wilson, Princeton, NJ: Princeton University Press, 1998.
8 J. Tyndall, *Fragments of Science: a Series of Essays, Addresses, and Reviews*, 10th imp., 1899, vol. 1, pp. 429–38, 422–8.
9 H.O. Sibum, 'Joule and Mayer', in J.L. Heilbron et al. (eds), *The Oxford Companion to the History of Modern Science*, Oxford: Oxford University Press, 2003, pp. 428–9.
10 J.P. Joule, *Scientific Papers*, London: Physical Society, 1884, vol. 1, pp. 298–328; 265–76.
11 C. Smith, 'Joule', in B. Lightman (ed.), *Dictionary of Nineteenth-century British Scientists*, Bristol: Thoemmes, 2004, vol. 2, p. 1109.
12 H. Chang and S.W. Yi, 'The Absolute and its Measurement: William Thomson on Temperature', *Annals of Science*, 62 (2005), 281–308.
13 D. Cahan (ed.), *Hermann von Helmholtz, and the Foundations of Nineteenth-century Science*, Berkeley, CA: California University Press, 1993, especially F. Bevilacqua, 'Helmholtz's *Erhaltung der Kraft*', pp. 291–333. See also the official biography, L. Koenigsberger, *Hermann von Helmholtz*, tr. F.A. Welby, intr. Lord Kelvin, Oxford: Oxford University Press, 1906 (reprint New York, Dover, 1965).
14 H. Helmholtz, *Popular Lectures on Scientific Subjects*, tr. E. Atkinson and J. Tyndall, London: Longman, 1873, pp. 153–96: quoted from pp. 153, 170, 193.
15 A. Warwick, *Masters of Theory: Cambridge and the Rise of Mathematical Physics*, Chicago, IL: Chicago University Press, 2003.
16 L. Campbell and W. Garnett, *The Life of James Clerk Maxwell*, new edn, London: Macmillan, 1884; B. Mahon, *The Man Who Changed Everything: the Life of James Clerk Maxwell*, Chichester: Wiley, 2003; and see Garnett's textbook, *Elementary Treatise on Heat*, 6th edn, Cambridge: Deighton Bell, 1893.
17 T.S. Kuhn, 'Energy Conservation as an Example of Simultaneous Discovery', *The Essential Tension: Selected Studies in Scientific Tradition and Change*, Chicago, IL: Chicago University Press, 1977, pp. 66–104.

18 L.P. Williams, *Album of Science: the Nineteenth Century*, New York, NY: Scribner's, 1978, p. 100.
19 G. Gooday, 'Sunspots, Weather, and the Unseen Universe: Balfour Stewart's Anti-materialist Representations of "Energy" in British Periodicals', in G. Cantor and S. Shuttleworth (eds), *Science Serialised*, Cambridge, MA: MIT Press, 2004, pp. 111–47.
20 B. Stewart and P.G. Tait, *The Unseen Universe*, 6th edn, London, 1876; *Paradoxical Philosophy*, London: Macmillan, 1878; the *Origin of Species* reached its sixth edition in 1872, thirteen years after the first.
21 P.C. Kjærgaard, 'Within the Bounds of Science: Redirecting Controversies to *Nature*', in L. Henson *et al.* (eds), *Culture and Science in the Nineteenth-century Media*, Aldershot: Ashgate, 2004, pp. 211–21, p. 214.
22 J.C. Maxwell, *Scientific Papers*, W.D. Niven (ed.), Cambridge: Cambridge University Press, 1890, vol. 2, pp. 756–62, pp. 757, 762.
23 G. Dawson, 'Victorian Periodicals and the Making of William Kingdom Clifford's Posthumous Reputation', G. Cantor and S. Shuttleworth (eds), *Science Serialized*, Cambridge, MA: MIT Press, 2004, pp. 259–84.
24 W.K. Clifford, *Lectures and Essays*, L. Stephen and F. Pollock (eds), London: Macmillan, 1879, vol. 1, pp. 228–253 – quotations from pp. 228, 253.
25 W.K. Clifford, *The Common Sense of the Exact Sciences*, 3rd edn, K. Pearson (ed.), London: Kegan Paul, Trench, Trübner, 1892.
26 W.H. Russell, *The Atlantic Telegraph* [1865], Newton Abbot: David and Charles, 1972; C. Smith and M.N. Wise, *Energy and Empire: a Biographical Study of Lord Kelvin*, Cambridge: Cambridge University Press, 1989, pp. 649–83.
27 G. Cookson and C.A. Hempstead, *A Victorian Scientist and Engineer: Fleeming Jenkin and the Birth of Electrical Engineering*, Aldershot: Ashgate, 2000; J. Simmons, *The Victorian Railway*, 2nd edn, London: Thames and Hudson, 1995, pp. 102–19.
28 W. Bagehot, *Collected Works*, N. St. J. Stevas (ed.), vol. 3, London: *The Economist*, 1968, p. 385.
29 *Turbinia* is on exhibition in the Museum of Science and Engineering in Newcastle, where she was built.
30 N. Rosenberg (ed.), *The American System of Manufactures: the Report of the Committee on the Machinery of the United States in 1855*, Edinburgh: Edinburgh University Press, 1969.
31 W. Russell, *The War*, 2 vols, London: Routledge, 1855–6.
32 W. Thomson, Lord Kelvin, *Baltimore Lectures on Molecular Dynamics and the Wave Theory of Light*, Cambridge: Cambridge University Press, 1904.
33 D.B. Wilson, *Kelvin and Stokes: a Comparative Study in Victorian Physics*, Bristol: Adam Hilger, 1987, pp. 129–80.
34 R. McCormmach, *Night Thoughts of a Classical Physicist*, Cambridge, MA: Harvard University Press, 1982.
35 A. Warwick, *Masters of Theory*, Chicago, IL: Chicago University Press, 2003, pp. 357–98.
36 O. Lodge, *Ether and Reality: a Series of Discourses on the Many Functions of the Ether of Space*, London: Hodder and Stoughton, 1925.
37 O. Lodge, 'The Work of Hertz', *Proceedings of the Royal Institution*, 14 (1893–5), 321–49.
38 W. Crookes, 'On Radiant Matter', *Nature*, 20 (1879), 419–23, 436–40; 'Electricity in Transitu; from Plenum to Vacuum', *Chemical News*, 63 (1891), 53–6, 68–70, 77–80, 89–93, 98–100, 112–14; reprinted in D.M. Knight (ed.), *Classical Scientific Papers; Chemistry*, London: Mills and Boon, 1970, pp. 89–98, 102–23; W.H. Brock, 'The Radiometer and its Lessons: William Carpenter versus William Crookes', in D.M. Knight and M.D. Eddy, *Science and Beliefs*, Aldershot: Ashgate, 2005, pp. 213–29.
39 J.M. Thomas and D. Phillips (eds), *Selections and Reflections: the Legacy of Sir Lawrence Bragg*, London: Royal Institution, 1990.

40 J.J. Thomson, 'Cathode Rays', *Proceedings of the Royal Institution*, 15 (1896–8), 419–32; 'Cathode Rays', *Philosophical Magazine*, 5th series, 44 (1897), 293–316.
41 Papers by Rutherford, J.J. Thomson and others are reprinted in facsimile in S. Wright (ed.), *Classical Scientific Papers: Physics*, London: Mills and Boon, 1964.
42 A. Warwick, *Masters of Theory*, Chicago, IL: Chicago University Press, 2003, pp. 399–409.
43 A.J. Balfour, in *Kelvin Centenary Oration and Addresses Commemorative*, London: Lund Humphries, 1924, p. 38.
44 W. Thomson, Lord Kelvin, 'Nineteenth-Century Clouds over the Dynamical Theory of Heat and Light', *Proceedings of the Royal Institution*, 16 (1899–1901), 363–97.
45 W. Bagehot, *Collected Works*, N. St.-J. Stevas (ed.), London: The Economist, 1968, vol. 3, p. 191.
46 See my 'Arthur James Balfour (1848–1930): Scientism and Scepticism', *Durham University Journal*, 87 (1995), 23–30; reprinted in my *Science in the Romantic Era*, Aldershot: Ashgate Variorum, 1998.
47 A.J. Balfour, *A Defence of Philosophic Doubt: Being an Essay on the Foundations of Belief*, London: Macmillan, 1879, p. 287.
48 *Papers Read Before the Synthetic Society, 1896–1908, and Written Comments thereon Circulated among the Members of the Society*, London: Spottiswoode for private circulation, 1909.
49 A.J. Balfour, *The Foundations of Belief: Being Notes Introductory to the Study of Theology*, 2nd edn, London: Macmillan, 1895, p. 31.
50 A.J. Balfour, 'Presidential Address', *Report of the British Association, Cambridge Meeting, 1904*, pp. 3–14; quotations from pp. 8, 9, 14.
51 G.G. Stokes, *Mathematical and Physical Papers*, vol. 2, Cambridge: Cambridge University Press, 1883, p. 97.
52 Lord Rayleigh, *Lord Balfour in his Relation to Science*, Cambridge: Cambridge University Press, 1930, p. 26.

13 Promoters and popularisers

1 N. Reingold, *Science in Nineteenth-century America: a Documentary History*, London: Macmillan, 1966.
2 H.E. Roscoe, *Life and Experiences*, London: Macmillan, 1906, pp. 281–313.
3 J. Lubbock, *Scientific Lectures*, London: Macmillan, 1879; *Ants, Bees and Wasps: a Record of Observations on the Habits of the Social Hymenoptera*, 16th edn, London: Kegan Paul, 1902; A.G. Duff (ed.), *The Life-work of Lord Avebury (Sir John Lubbock), 1834–1913*, London: Watts, 1924.
4 L. Huxley (ed.), *The Life and Letters of Sir Joseph Dalton Hooker*, London: Murray, 1918, vol. 2, p. 71.
5 M.P. Crosland, *Science Under Control: the French Academy of Sciences 1795–1914*, Cambridge: Cambridge University Press, 1992, pp. 187–92, 300–30, 396–441.
6 N.M. Brooks, 'The Evolution of Chemistry in Russia during the 18th and 19th centuries', in D. Knight and H. Kragh (eds), *The Making of the Chemist: the Social History of Chemistry in Europe*, Cambridge: Cambridge University Press, 1992, pp. 163–76.
7 M. Shortland and R. Yeo (eds), *Telling Lives in Science: Essays on Scientific Biography*, Cambridge: Cambridge University Press, 1996, e.g. pp. 12, 103.
8 P.J. Bowler, 'From Science to the Popularisation of Science: the career of J. Arthur Thomson', in D.M. Knight and M.D. Eddy (eds), *Science and Beliefs: from Natural Philosophy to Natural Science, 1700–1900*, Aldershot: Ashgate, 2005, pp. 231–48.
9 D.M. Knight, *Science and Spirituality: the Volatile Connection*, London: Routledge, 2004, pp. 151–66.
10 J.C. Maxwell, *The Scientific Papers*, W.D. Niven (ed.), Cambridge: Cambridge University Press, 1890, vol. 2, pp. 445–84, 485–91.

11 K.A. Neeley, *Mary Somerville: Science, Illumination, and the Female Mind*, Cambridge: Cambridge University Press, 2001.
12 On him and others, see B. Lightman (ed.), *Dictionary of 19th-century British Scientists*, Bristol: Thoemmes, 2004.
13 A. Lundgren and B. Bensaude-Vincent, *Communicating Chemistry: Textbooks and their Audiences*, Canton, MA: Science History Publications, 2000.
14 E. Turner, *Elements of Chemistry*, 3rd edn, Edinburgh: Maclachlan, 1831; J.F. Daniell, *An Introduction to the Study of Chemical Philosophy: a Preparatory View of the Forces which Concur to the Production of Chemical Phenomena*, London: Parker, 1839.
15 J. Issit, 'Jeremiah Joyce', in B. Lightman (ed.), *Dictionary of Nineteenth-century British Scientists*, Bristol: Thoemmes, 2004, vol. 2, pp. 111–14; his book on Joyce is in press, Aldershot: Ashgate.
16 J. Guy, *Elements of Astronomy: Familiarly Explaining the General Phœnomena of the Heavenly Bodies and the Theory of the Tides*, 2nd edn, London: Baldwin, Cradock and Joy, 1821, pp. vii, viii, 138, 6.
17 A. Fyfe, 'Expertise and Christianity: High Standards Versus the Free Market in Popular Publishing', in D.M. Knight and M.D. Eddy (eds), *Science and Beliefs*, Aldershot: Ashgate, 2005, pp. 113–26.
18 J. Taylor, *Anecdotes of Remarkable Insects; Selected from Natural History, and Interspersed with Poetry*, London: Baldwin, Cradock and Joy, 1817.
19 *Pliny's Natural History: a Selection from Philemon Holland's Translation*, J. Newsome (ed.), Oxford: Oxford University Press, 1964; *Selections from the History of the World, Commonly Called the Natural History of C. Plinius Secundus*, tr. P. Holland, P.Turner (ed.), London: Centaur, 1962.
20 A. Ashfield (ed.), *Romantic Women Poets, 1770–1838: an Anthology*, Manchester: Manchester University Press, 1995; J. Breen (ed.), *Women Romantic Poets, 1785–1832: an Anthology*, London: Dent, 1992.
21 W. Kirby and W. Spence, *An Introduction to Entomology: or Elements of the Natural History of Insects*, London: Longman, vol. 1, 4th edn, 1822; vol. 2, 3rd edn, 1823; vol. 3, 1826; vol. 4, 1826.
22 W. Curtis, *A Short History of the Brown-tail Moth* [1782], W.T. Stearn and D.S. Fletcher (eds), London: Curwen, 1969.
23 V.M. Holt, *Why Not Eat Insects?* [1885], Faringdon: Classey, 1978.
24 A. Pratt, *The Poisonous, Noxious, and Suspected Plants of our Fields and Woods*, London: SPCK, n.d., pp. vii, 144, 152.
25 J.G. Wood, *Common Objects of the Microscope*, London: Routledge, 1866; *Homes Without Hands*, London: Longman, 1866; R.A. Gilbert, 'J.G. Wood', in B. Lightmen (ed.), *Dictionary of Nineteenth-century British Scientists*, Bristol: Thoemmes, 2004, vol. 4, pp. 2193–6.
26 F.O. Morris, *A History of British Butterflies*, 8th edn, London: Nimmo, 1895, p. vii.
27 W.H. Hudson, *Birds in London* [1898], R. Fitter (ed.), Newton Abbot: David and Charles, 1969.
28 C. Knight, *The Old Printer and the Modern Press*, London: Murray, 1854.
29 *A Description and History of Vegetable Substances, Used in the Arts, and in Domestic Economy: Timber Trees, Fruits*, London: Knight, 1829.
30 *Popular Philosophy: or the Book of Nature Laid Open upon Christian Principles*, by the Editor of the Cheap Magazine and Monthly Monitor, Dunbar: Miller, 1826.
31 H. Lonsdale, *A Sketch of the Life and Writings of Robert Knox, the Anatomist*, London: Macmillan, 1870.
32 K.G.V. Smith, 'F.O. Morris', in B. Lightman (ed.), *Dictionary of Nineteenth-century British Scientists*, Bristol: Thoemmes, 2004, vol. 3, pp. 1427–9.
33 H. Litchfield (ed.), *Emma Darwin: a Century of Family Letters, 1792–1896*, London: Murray, 1915, vol. 2, pp. 219–21.
34 A. Guerrini, 'Animal Care and Experimentation', in J.J. Heilbron (ed.), *The Oxford*

Companion to the History of Modern Science, Oxford: Oxford University Press, 2003, pp. 27–9.
35 H. Litchfield (ed.), *Emma Darwin: a Century of Family Letters, 1792–1896*, London: Murray, 1915, vol. 2, p. 87.
36 H.E. Roscoe, *Life and Experiences*, London: Macmillan, 1906, pp. 314–36.
37 G.L. Geison, *The Private Science of Louis Pasteur*, Princeton, NJ: Princeton University Press, 1995.
38 A.R. Wallace, *My Life: a Record of Events and Opinions*, London: Chapman and Hall, 1908, p. 330.
39 M. Bulmer, 'The Theory of Evolution of Alfred Russel Wallace', *Notes and Records of the Royal Society*, 59 (2005), 125–36.
40 T. Cooper, *Evolution, the Stone Book, and the Mosaic Record of Creation*, London: Hodder and Stoughton, 1878.
41 Galileo, *Dialogue Concerning the Two Chief World Systems* [1632], ed. and tr. S. Drake, Berkeley, CA: California University Press, 1962.
42 B. le B. De Fontenelle, *Entretiens sur la pluralité des mondes*, R. Shackleton (ed.), Oxford: Oxford University Press, 1955; *A Week's Conversation on the Plurality of Worlds*, tr. A. Behn *et al.*, London: Bettesworth, 1737.
43 R. Boyle, *The Sceptical Chymist* [1661], intr. M.M. Patison Muir, London: Dent, 1911; H. Knight, 'Rearranging 17th-century Natural History into Natural Philosophy: 18th-century Editions of Boyle's Works', in D.M. Knight and M.D. Eddy (eds), *Science and Beliefs*, Aldershot: Ashgate, 2005, pp. 31–42.
44 B. Brodie, *Psychological Inquiries: in a Series of Essays, Intended to Illustrate the Mutual Relations of the Physical Organization and the Mental Faculties*, 2nd edn, London: Longman, 1855; and see B.C. Brodie, *Autobiography*, 2nd edn, London: Longman, 1865, esp. pp. 88–103.
45 J. Kepler, *The Dream*, tr. P.F. Kirkwood, J. Lear (ed.), Berkeley, CA: California University Press, 1965.
46 C. de Bergerac, *Other Worlds: the Comical History of the States and Empires of the Moon and the Sun*, tr. G. Strachan, Oxford: Oxford University Press, 1965.
47 M. Shelley, *Frankenstein: or, the Modern Prometheus* [1818], D.L. Macdonald and K. Scherf (eds), 2nd edn, Peterborough, Ontario: Broadview, 1999.
48 H.G. Wells, *The Time Machine* [1895], J. Lawton (ed.), London: Dent, 1995.
49 M. Hawkins, *Social Darwinism in European and American Thought, 1860–1945: Nature as Model, and Nature as Threat*, Cambridge: Cambridge University Press, 1997.
50 D. Pick, *Faces of Degeneration: a European Disorder, c.1848–c.1918*, Cambridge: Cambridge University Press, 1989; A. Kuper, 'Incest, Cousin Marriage, and the Origin of the Human Sciences in 19th-century England', *Past and Present*, 174 (2002), 158–83.
51 C. Lombroso and W. Ferrero, *The Female Offender*, intr. W.D. Morrison, London: Fisher Unwin, 1895.
52 A. Owen, *The Place of Enchantment: British Occultism and the Culture of the Modern*, Chicago, IL: Chicago University Press, 2004.
53 S. Petruccioli (ed.), *Storia della Scienza*, vol. 8, *La Seconda Rivoluzione Scientifica*, Rome: Enciclopedia Italiana, 2004.
54 L. Huxley (ed.), *The Life and Letters of Sir Joseph Dalton Hooker*, London: Murray, 1918, vol. 2, pp. 159–77.
55 P.J.T. Morris (ed.), *From Classical to Modern Chemistry: the Instrumental Revolution*, London: Royal Society of Chemistry, 2002.
56 J. Morrell, *John Phillips and the Business of Victorian Science*, Aldershot: Ashgate, 2005, pp. 223–8.
57 J. Krige and D. Pestre (eds), *Science in the Twentieth Century*, Amsterdam: Harwood, 1997.
58 C.G. Gaither and A.E.C. Gaither (eds), *Chemically Speaking: a Dictionary of Quotations*, Bristol: Institute of Physics, 2002, p. 118.

Index

Academy of Sciences, Paris 3–4, 18, 47, 84, 91, 95, 138, 170, 172
Acton, Hannah 26
Africa 117–18
Agassiz, Louis 23, 24, 74
agnosticism 2, 42, 43, 161
agriculture 37, 76–7, 79–82, 86–7
Albert, Prince Consort 40, 87, 89, 100
alchemy 46–7, 63
algebra 48–9, 57, 75
anatomy 29–30, 94, 96–7, 189
Anecdotes of Remarkable Insects 186
Anglican Church, dissenters from 13–28
animal experiments *see* vivisection
Annals of Philosophy 141–2, 144
Annals of Science 139
Anthropological Society 147
applied science 4–5, 33–4, 76–90, 133–4
Architecture of the Heavens 67
Armstrong, William 104, 174
Ashmolean Museum, Oxford 102–3
Asia 111, 112, 115, 118
astronomy 17, 21–2, 56, 67, 106–8
Athenaeum Club 85, 182
atomic theory 128–30, 141–2
Australia 86, 108–10

B.A.A.S. 8, 41, 59–60, 86, 97, 102, 126, 132, 159, 169, 175, 180, 181
Babbage, Charles 56
Bacon, Francis 153, 154, 155, 156
Bagehot, Walter 178–9
Balfour, Arthur 128, 137, 179–81
Banks, Joseph 11, 14–15, 33, 34–5, 52, 78–9, 85, 98, 107–8, 113
Bates, Henry Walter 114–15, 116
Beddoes, Thomas 31, 51
Bell, Charles 21, 189
Berthollet, Claude-Louis 38–9, 47
Berzelius, Jacob 64–5, 86, 129
Bewick, Thomas 69–70
Birmingham 32

Board of Agriculture 80
Botanic Garden, The (1791) 15
botanic gardens 4, 91–5, 98–100
botany 15–16, 34, 44–5, 72, 109
Bourne, John 68–9
Boyle, Robert 16, 130, 192
Brande, William Thomas 40, 65, 140, 184
Bridgewater Treatises 20–4, 156
British Association for the Advancement of Science (B.A.A.S.) *see* B.A.A.S.
British Museum 93, 96, 100, 154, 165
British Union for the Abolition of Vivisection 190
Brown, Robert 109
Brunel, Isambard 69
Buckland, Frank 96
Buckland, William 21, 22–4, 26, 73, 85, 94

Cabinet Cyclopaedia 56, 70–1, 156
Cambridge University 30–1, 77, 120, 156, 157, 161, 164
Canada 107, 112–13
caricature 61, 73, 74
Carnot, Sadi 78, 167, 170
Celestial Scenery (1839) 26
Century of Progress exhibition, Chicago (1876) 103
Chambers, Robert 24–5, 67
Chandler, Richard 117
Chemical Catechism 6, 26, 184
Chemical History of a Candle, The 40
Chemical Manipulation 62, 113
Chemical News 8–9, 149
Chemical Society 60, 129, 149
chemical symbols 63, 75, 86
Chemist, The 144
chemistry 26–7, 30–1, 35–40, 46–61, 63–7, 79–84
Christian Philosopher, The (1823) 26
Church Scientific 183
Clanny, William 83
classical physics 167–81

Index

clergy 82
Clifford, William Kingdom 137, 173
clocks 16–17
coal mining 69; *see also* miners' lamp
Cobb, Frances Power 105, 189–90
Cole, Henry 87, 89, 100, 164
Coleridge, Samuel Taylor 1, 51, 53, 120–1, 155–6, 157
College of Physical Science, Newcastle-upon-Tyne 164
College of Surgeons 100
Collins, Edward James Mortimer 123–4
Common Objects of the Microscope 187
Common Sense of the Exact Sciences (1885) 173
Comte, Auguste 122–3, 125, 161–2, 167
Conservatoire des Arts et Metiers, France 93
Consolations in Travel, or the Last Days of a Philosopher (1830) 20, 53, 155–6
Conversations on Chemistry 6, 38, 62, 65, 184, 185
Cook, James 10, 11, 103, 107–8, 109, 113–14
Cooper, Thomas 191
Copley Medal 22, 37, 39, 79, 111, 169, 172
Crookes, William 8–9, 59, 133, 147, 148–9, 160, 176
Cruelty to Animals Act (1876) 190
Crystal Palace 68, 88, 99, 100, 103, 118, 174
Curie, Marie 176
Curtis's Botanical Magazine 99
Cuvier, Georges 22, 92, 93–4

Dalton, John 32, 48, 86, 128–9, 141
Daniell, J.F. 184
Darwin, Charles 11–12, 44, 73, 97, 99, 111–12, 115–16, 122, 132–3, 137–8, 158, 159–60, 189, 190, 194
Darwin, Erasmus 15–16, 44–6, 48, 49, 53
Davy, Sir Humphry 1, 4–5, 18–20, 33, 35–41, 51, 53–6, 65, 79–80, 82–4, 94, 119–20, 125, 143, 146, 154–6, 168, 194–5; *see also* miners' lamp
Defence of Philosophic Doubt, A 179
Deleuze, Joseph 93–4
demonstration experiments 31, 36, 37–8, 65–6
Denmark 159
Dick, Thomas 26
Diquisitions on Matter and Spirit (1783) 14
display 91–105
dissection 29–30, 96–7, 189
Dissenting Academies 31, 77–8

Ecole Polytechnique, France 92
Edinburgh Review 137
education 6–8, 31–2, 42–3, 66, 89–90, 119–20, 123, 126, 163–6
Einstein, Albert 178, 180
electrochemistry 6, 84

electromagnetism 159
Elements of Astronomy 185–6
Elements of Chemistry (1789) 44, 47, 63
Elements of Physiophilosophy 160
embryology 25
encyclopaedias 68, 155, 183
energy conservation 130–2
engineering drawings 68
engravings 63–4, 69–70, 72, 188
evangelicals 26
Evidences 17
evolution 132–3; *see also* natural selection
Evolution, the Stone Book and the Mosaic Record of Creation (1878) 191
Exeter Change 59–60
exhibitions 68–9, 87–9, 100–3
experimental chemistry 63–7
explorers *see* scientific exploration
explosive gases 82–4

Faraday, Michael 1, 28, 38, 39, 40, 54–5, 62, 67, 84–5, 122, 143–4, 159, 176, 184
Female Offender, The 194
FitzGerald, George 175
Flinders, Matthew 109, 113
Flower, William Henry 97–8
foreign journals 140
Forster George 108
Foster Henry 111
Foucault, Léon 172
Foundations of Belief, The 179–80
Fourcroy, Antoine Francois 4, 47
Fownes, George 26–7
Fragments of Science 126, 169
France 3–4, 18, 86, 91–6, 103, 109, 156, 164, 165, 167–8
Frankenstein (1818) 9, 26, 192–3
Franklin, John 112–13

Garnett, Thomas 35
Genesis 10, 23, 26, 43
Gentleman's Magazine 147
Geological Society 142
geology 22–3, 24, 94–5
geometry 48, 64, 75
George III, King 66, 98
Germany 6–7, 32, 84, 108, 140, 156, 159–61, 164, 165
Gosse, Philip 28, 86
Gould, John 70, 72
graphical presentations 74–5
grave robbing 189
Great Exhibition: (1851) 68–9, 87–9, 103; (1862) 100
Greenwich 106, 107

Halley, Edmond 106, 107
Hardwicke's Science Gossip 150–1

230 Index

Hawkesworth, Dr John 108
Helmholtz, Hermann 125, 169–71
Henry, Joseph 168
Herschel, John 56–8, 85, 112, 113, 122, 156–7
Herschel, William 67, 75, 168
Hertz, Heinrich 176
History of European Thought in the Nineteenth Century 3
HMS *Beagle* 103, 111, 112
HMS *Endeavour* 79, 107–8, 109
Holt, Vincent 186
Hooker, William 98–9, 182
Horticultural Society 95
Humboldt, Alexander von 11, 74, 110, 112, 114
Hume, David 162
Huxley, Thomas Henry 9–10, 28, 41–3, 72, 86, 97, 124, 125, 132, 137, 151, 152, 153–4, 156, 160–3, 182, 183
hydrography 111

illustration 12, 23–4, 62–77, 110, 168–9, 188
imagination 119–34
Imperial College 102
Imperial Institute 89, 118
In Memoriam 28
India 112, 118
Inductivism 124–5
industrialisation 18
international loan exhibition, scientific apparatus (1876) 100–2

Japan 111
Jardin des Plantes, France 4, 91–5, 98
Jenner, Edward 190–1
Joule, James 32, 169
Journal of Natural Philosophy, Chemistry and the Arts see Nicholson's Journal
journals 6, 8–9, 11, 81–2, 108–9, 135–52
Joyce, Jeremiah 5, 62, 185
Jung, Carl Gustav 46–7

Kant, Immanuel 21, 31, 43
Kepler, Johannes 192
Kew 98–100, 112, 165, 195
Kirby, William 21, 186
Knowles, James 9, 137
Kuhn, Thomas 49–50, 132, 171

laboratories 64–7
Lamarck, J.B. 92
Laplace, Pierre-Simon 21–2, 67, 156
Lardner, Dionysius 56, 70–1, 156
Latin America 110
Lavoisier, Antoine 9, 18, 38, 44, 47–9, 50, 55, 58, 60–1, 63–4, 75, 91, 140, 167–8, 184
Lawrence, Honoria 118
Lear, Edward 71–2

lectures 29–43
Lectures and Essays 173
Liebig, Justus 6–7, 27, 66, 184
Linnaeus, Charles 15, 184–5
Literary and Philosophical Society, Manchester 32, 169
lithography 24, 69–70
Liverpool 32
Lockyer, Norman 9, 149, 163
Lombroso, Cæsar 194
London Missionary Society 111
London School Board 42–3, 152, 163
London Zoo 95–6, 97
Loves of the Plants, The 15
Lunar Society of Birmingham 15, 32, 44, 78
Luttrell, Henry 39
Lyell, Charles 22–3, 51, 132, 158
Lyrical Ballads 51

Macquer, Pierre Joseph 47, 48, 49
Magazine of Natural History, The 144–7, 148–9
magnetic observations 114
Manchester 32
Mansel, Henry 43, 161
Manual of Scientific Enquiry (1849) 12, 113
mapping 11
Marcet, Jane 6, 38, 62, 65
Marshall, William 80
Martineau, Harriet 124, 161
mathematical instruments 66–7
Maxwell, James Clerk 58–9, 130–1, 171, 172–3, 178
Mayer, Julius Robert 169
Mechanics Magazine, The 144
Mechanics' Institutes 26, 90
medical science 189–91
Melincourt 15
Mendeleev, Dmitri 148, 177, 184–5
Michelson, Albert 175
Midsummer-Night's Dream 121–2
military reviews 103–4
Mill, John Stuart 5–6, 62, 122, 157–8, 159
Miller, Hugh 24
mineralogy 64–5
miners' lamp 4–5, 36, 39, 79, 82–4
Moral Philosophy 17
Morley, Edward 175
Morris, Francis Orpen 187–8, 189
Murchison, Sir Roderick 20, 117
Museum of Practical Geology 41–2
museums 41–2, 89–90, 91–105, 154, 165

National Anti-Vaccination League 191
National Anti-Vivisection Society 189–90
natural history 23–4, 44–7, 70–2, 91–100, 106, 186–8

Index

Natural History Museum 89, 96–8, 99, 102, 154
natural philosophers 4–5, 122
natural selection 11–12, 15–16, 28, 97, 114–16, 131–2, 191
natural theology 13–28, 186
Natural Theology (1802) 12, 16–17
Nature 9, 59, 116, 149–50, 172–3
Naturphilosophie 159–60
New Zealand 108
Newcastle-upon-Tyne 32–3, 164
Newcomen, Thomas 133
Newton, Isaac 2, 17, 44, 50, 56, 63, 64, 138, 156
Nicholson, William 6, 49, 50, 140–1
Nicholson's Journal 140–1, 144, 151
Nineteenth Century, The 9, 137
Nobel, Alfred 174
Normal School 89, 154
normal science 49–50, 132
North, Marianne 99, 118

observatories 106, 107
Odling, William 148
Oersted, Hans Christian 27, 57, 158–60, 168
Oken, Lorenz 28, 84, 160
Origin of Species (1859) 11, 44, 73, 117, 122, 131, 132, 138, 156, 158
Owen, Richard 96–7, 98, 99, 132
Owen, William 111
Oxford University 22–3, 30–1, 43, 77, 120, 158, 161, 164

palaeontology 94–5
Paley, William 16–18, 21–2
Pantheism 53–4, 125
Paradoxical Philosophy (1878) 172–3
Paris Exposition (1900) 103
Parkes, Samuel 6, 26, 184
parody 52–3, 58–61
Parsons, Charles 104, 174
Pasteur, Louis 104
Pauling, Linus 196
Peacock, George 56, 57
Periodic Table 129, 148, 185
Phillip, Captain Arthur 108–9
Phillips, John 41, 85
Philosophical Magazine, The 6, 141, 142, 144
Philosophical Transactions 5, 35, 138–40, 144, 163, 183
philosophy 49–50, 52–3, 123–4, 153–6, 157–9, 161–2
photography 64, 73–4
phrenology 10, 41, 74, 103
physics 130–2, 167–81
physiology 189
Planck, Max 178
Playfair, William 75, 87
Pneumatic Institution 51

poetry 15–16, 44–61
polar expeditions 114
Popular Philosophy 188
Positivism 123–4, 161–2
pottery manufacture 78
Pratt, Anne 186–7
Preliminary Discourse 56, 122, 156
Priestley, Joseph 14–15, 17–18, 46, 50–1, 74–5, 77–8, 168
Principles of Geology (1830) 23
printing methods 135
Proceedings of the Royal Society 139
Prout, William 21, 141–2, 177
provincial cities 32, 41
provincial museums 102–3
psychology 162
public schools 66
Punch 61, 136, 137

Quarterly Journal of Science 8, 147–9
Quarterly Journal of Science, Literature and the Arts 143–4
Quarterly Review 138, 156–7

Raffles, Sir Stamford 95
railways 68–9, 88, 110–11
Records of General Science 142
Red Lions 59
Reform Bill (1832) 21
Register of Arts and Sciences 142–3
Reichenbach, Karl von 10
religious dissenters 12, 13–28, 77–8
religious philosophies 157–9
Religious Tract Society 26
reviews 137–8, 172–3
revolutions (1848) 25
Roget, Peter Mark 21
Röntgen, Wilhelm Konrad 176
Royal Agricultural Society 86–7
Royal College of Art 102
Royal College of Chemistry 89, 134
Royal Commissions 43, 66, 163–4, 182, 191
Royal Geographical Society 11, 115, 117
Royal Institution 8, 18–19, 26, 33–7, 40–1, 65–6, 79, 93, 179
Royal Medal 39, 129
Royal Microscopical Society 150
Royal Navy 103–14, 161
Royal Society 34–5, 37, 41–2, 81–2, 114, 138–40, 153; Presidency 11, 19, 34, 39–40, 51, 78, 85, 95, 99, 103, 149; *see also* Bridgewater Treatises; Copley Medal; *Philosophical Transactions*; *Proceedings of the Royal Society*
Royal Society Catalogue of Scientific Papers 140
Rutherford, Ernest 177, 181

Samuelson, James 147
Sceptical Chymist 192
School of Mines 89, 90, 154
science fiction 9, 12, 192–4
science gossip 135–52
Science Gossip [Hardwicke's] 150–1
Science Museum 89, 102
science promoters/popularisers 182–96
science wars 1, 2, 86, 132
scientific apparatus 100–2
Scientific Dialogues (1807) 5, 62
scientific exploration 10–12, 79, 99, 103, 106–18
scientific method 153–66
scientific papers 138–40, 182–3
Scientific Periodicals Project 136
scientific training 6–8, 119–20, 133–4, 153–4
scientific understanding 1–12
scientists, illustrations of 72–4
Scotland 32
Scott, David 67
Scott, John 92–3
Shelley, Mary 26, 192–3
Shelley, Percy Bysshe 51
slave trade 111, 117
Snow, Charles 7, 131
social classes, integration of 41–2
Social Darwinism 194
Society for Promoting Christian Knowledge 187
Society for Psychical Research 10
Society for the Encouragement of Agriculture, Arts, Manufactures and Commerce 81
Society for the Protection of Birds 116
Society of Arts 77, 87
Society of Dilettanti 117
Soddy, Frederick 177
Somerville, Mary 8, 90, 183–4
Soul in Nature, The 27, 57, 158–9
South America 107, 111–12, 114–15, 116
Spain 13
specialists 2, 7, 8, 122
spiritualism 10, 194
steam engines 78, 133–4
Stephenson, George 4, 83
Stokes, George Gabriel 157, 163
Swainson, William 70, 71, 72, 146
Swayne, Revd G. 144

technical language 58
telepathy 10
Temple of Nature 45, 53
Tennyson, Alfred Lord 28, 51–2, 58–60
textbooks 36–7, 66, 183–8
textile manufacturing 77
theology 13–29, 161, 186

Theory of Life 155
thermodynamics 130–1, 133–4
Thompson, Benjamin 33–4, 35–6
Thomson, J. Arthur 182–3
Thomson, J.J. 176–7, 181
Thomson, Thomas 5–6, 66, 141–2
Thomson, William 67, 129, 169, 170, 171, 173–4, 175
Tilloch, Alexander 6, 141
Time Machine 9, 132, 193–4
Tractarians 158
trades 76–7
translation 50
travel 106–18
Turbinia 104, 174
Tyndall, John 125–8, 165–6, 169–70, 172

Unitarians 14–15, 16
United States of America 86, 110–11, 164
universities 6–7, 66, 89–90, 134, 164–5, 184; *see also* Cambridge University; Oxford University
Ure, Dr Andrew 29–30
utilitarianism 17–18

vaccination 104, 190–1
Verne, Jules 193
Vestiges of the Natural History of Creation (1844) 24–5, 51–2, 67, 132, 160
Victoria and Albert Museum 89, 100, 102
Victoria, Queen 40
vivisection, condemnation of 10, 104–5, 189–90

Wallace, Alfred Russel 114–16
warships 84
Watson, Richard 30–1
Watt, James 78
weaponry 103–4
Wells, Herbert George 9, 132, 193–4
Whewell, William 7–8, 21, 22, 56, 87–8, 122, 156–8, 184
Why not Eat Insects (1885) 186
Wilson, George 27, 89
Wollaston, William Hyde 64, 65, 129, 140, 154–5, 168
women 7, 93–4, 106, 118, 189
Wood, John George 187
Wordsworth, William 51, 53, 121–2, 126
Wurtz, Adolphe 60–1

X-club 125, 165

Yorkshire Philosophical Society 85–6
Young, Arthur 80

zoological gardens 91, 95–7